J. Richard Gott is a profes............. ....,.... ...-ences at Princeton University. For fourteen years he served as the chair of the judges of the US National Westinghouse and Intel Science Talent Search, the premier science competition for high-school students. A recipient of the President's Award for Distinguished Teaching at Princeton, Gott has written on time travel for *Time* and on other topics for *Scientific American, New Scientist* and *American Scientist.* He lives in Princeton, New Jersey.

TIME TRAVEL
IN EINSTEIN'S
UNIVERSE

The Physical Possibilities of
Travel Through Time

J. RICHARD GOTT

WEIDENFELD & NICOLSON

First published in Great Britain in 2001
by Weidenfeld & Nicolson
This paperback edition published in 2005
by Weidenfeld & Nicolson,
an imprint of The Orion Publishing Group Ltd,
Carmelite House, 50 Victoria Embankment
London EC4Y 0DZ

An Hachette UK company

First published in the USA in 2001
by Houghton Mifflin

11

Copyright © J. Richard Gott III 2001

A CIP catalogue record for this book
is available from the British Library.

ISBN 978-0-7538-1349-2

Printed and bound in Great Britain by
Clays Ltd, Elcograf S.p.A.

MIX
Paper from
responsible sources
FSC
www.fsc.org   FSC® C104740

www.orionbooks.co.uk

Dedicated to—

My mother and father, wife and daughter

—my past, present, and future

CONTENTS

ACKNOWLEDGMENTS

First and foremost, I thank my lovely wife, Lucy, my soul mate—
for believing. Since Lucy is one of the smartest people around
(summa cum laude at Princeton), I always take her advice very seriously! For this book she has added her considerable professional
skills as an editor and writer to help me produce a much improved
manuscript. To my daughter, Elizabeth—one could not hope for a
better daughter. In addition to lighting up our lives, she has taken
time from her stellar high school career to help me as well, sometimes by creating a computer system, but more often by helping
me find the right visual aids to explain physics concepts. She found
the cute, chubby space shuttle I used to show how one might circle
two cosmic strings (pictured in *Time*), and she discovered the tiny,
flag-waving astronaut for me to drop into a funnel to illustrate the
properties of black holes (for *The McNeil-Lehrer Newshour*). To my
mother and father, Marjorie C. Gott and Dr. John Richard Gott, Jr.,
I offer my thanks for their support over the years, including the
way my mother cheerfully took me to countless Astronomical
League conventions and science fairs during my high school years.

I would like to thank especially Laura van Dam, my wonderful
editor at Houghton Mifflin, who first came to me with the idea that
I should write a book on time travel. Her enthusiasm, incisive judg-
ment, and abundant editorial talent have made working with her a
joy. I also thank Liz Duvall, Susanna Brougham, and Lisa Diercks
for gracious help during the production process.

For turning my sketches into beautiful line drawings and graphics, I thank JoAnn Boscarino and Li-Xin Li, respectively. Some of the diagrams were created with the Mathematica program, Claris-Works, or Design It! 3-D.

Charles Allen (president of the Astronomical League) and Neil de Grasse Tyson (director of the Hayden Planetarium) read the entire manuscript. Their feedback has been essential; more so, their friendship over the years. Jonathan Simon and Li-Xin Li read selected chapters and offered useful comments. I also benefited from comments by Jeremy Goodman, Suketu Bhavsar, Deborah Freedman, Jim Gunn, Frank Summers, Douglas Heggie, Ed Jenkins, Michael Hart, Matthew Headrick, Jim Peebles, Bharat Ratra, and Martin Rees.

I am grateful to all my teachers (from my high school math teacher, Ruth Pardon, to my thesis adviser, Lyman Spitzer) and my many colleagues, who include my students. Special thanks to Li-Xin Li whose collaboration on our research described in Chapter 4 has been pivotal. Figure 27 is from our 1998 *Physical Review* paper "Can the Universe Create Itself?" I would like to thank George Gamow and Charles Misner, Kip Thorne, and John Wheeler, whose books have been a source of inspiration to me; Hugh Downs, for many lively cosmology dinners; and Carl Sagan and again Kip Thorne, whose interest in my work I have greatly appreciated. I thank Dorothy Schriver and all the people I've known at Science Service; my mother-in-law, Virginia Pollard; and Drs. William Barton and Alexander Vukasin. I also wish to acknowledge the science writers who have done excellent pieces on my work: Timothy Ferris, Michael Lemonick, Sharon Begley, James Gleick, Malcolm Browne, Marcus Chown, Ellie Boettinger, Kitta MacPherson, Gero von Boehm, Joel Achenbach, Marcia Bartusiak, Mitchell Waldrop, and Rachel Silverman. Because of science writers like these, the wide panoply of scientific endeavor is opened to all. I hope this book will add to this in some small measure.

Finally, I salute Albert Einstein, whose ideas challenge us still.

PREFACE

The neighborhood children think I have a time machine in my garage. Even my colleagues sometimes behave as if I have one. Astrophysicist Tod Lauer once sent me a formal letter inviting me to Kitt Peak National Observatory to give a talk on time travel. He sent this invitation six months *after* I had already given the talk. The invitation explained that since I was an expert in time travel, I should presumably have no trouble in returning to the past to make the appearance. On another occasion, at a cosmology conference in California, I happened to wear a turquoise sports jacket —which I imagined might fit in nicely with the California ambiance. Bob Kirshner, then chair of Harvard's astronomy department, came up to me and said, "Richard, this is the 'Coat of the Future'; you must have gotten this in the future and brought it back, because this color hasn't been invented yet!" Since then, I've always worn this coat when giving talks on time travel.

Time travel is certainly one of the most fun topics in physics, but it has a serious side as well. I have received calls from people who want to know about recent developments in time travel because they wish to return to the past to rescue a loved one who died under tragic circumstances. I treat such calls with great seriousness. I have written this book partly to answer such questions. One reason that time travel is so fascinating is that we have such a great desire to do it.

Physicists like me who are investigating time travel are not currently at the point of taking out patents on a time machine.

But we are investigating whether building one is possible in principle, under the laws of physics. It's a high-stakes game played by some of the brightest people in the world: Einstein showed that time travel to the future is possible and started the discussion. Kurt Gödel, Kip Thorne, and Stephen Hawking have each been interested in the question of whether time travel to the past is possible. The answer to that question would both give new insights into how the universe works and possibly some clues as to how it began.

This book is a personal story, not a history of science. Imagine me as your guide, taking you to the summit of Mount Everest. The climb is sometimes challenging, sometimes easy, but I promise that we will ascend by the easiest possible route. It's a path of ideas I know well, having marked some of the trail myself. Along the way, we will intersect the work of many of my colleagues. I have mentioned many of them to give you a fair idea of the other trailblazers of this terrain. Some contributions are emphasized and others briefly noted, in or out of historical sequence, as they play into telling my story. To those whose work I've not mentioned—though it may be equally important but following a different route up the mountain—I apologize in advance.

We start our journey at base camp: the dream of time travel itself and the pathbreaking science fiction of H. G. Wells.

TIME TRAVEL IN EINSTEIN'S UNIVERSE

1 DREAMING OF

TIME TRAVEL

> Man . . . can go up against gravitation in a balloon, and
> why should he not hope that ultimately he may be able
> to stop or accelerate his drift along the Time-Dimension,
> or even turn about and travel the other way.
> — H. G. WELLS, *THE TIME MACHINE*, 1895

WHAT WOULD YOU DO WITH A TIME MACHINE?

No idea from science fiction has captured the human imagination as much as time travel. What would you do if you had a time machine? You might go to the future and take a vacation in the twenty-third century. You might bring back a cure for cancer.

Then again, you might return to the past to rescue a lost

loved one. You could kill Hitler and prevent World War II or book passage on the *Titanic* to warn the captain about the iceberg. But what if the captain ignored your warning, as he ignored all the other warnings about icebergs that he received, so that the great ship sank after all? In other words, would time travel let you change the past? The notion of time travel to the past can suggest paradoxes. What if, on a trip to the past, you accidentally killed your grandmother before she gave birth to your mother?

Even if changing the past is impossible, going there might still be very interesting. Even if you could not change history from the course we know it took, you still could participate in shaping that history. For example, you might go back in time to help the Allies win the Battle of the Bulge in World War II. People love to reenact Civil War battles—what if it were possible to participate in the real thing? Selecting a battle won by your side would give you the thrill of joining in the experience as well as the secure feeling of knowing the outcome. In fact, it might turn out that, in the end, the tide of battle was turned by tourists from the future. Indeed, people who have been far ahead of their time in their thinking, such as Jules Verne and Leonardo da Vinci, have sometimes been accused of being time travelers.

If you chose to embark on time travel, you could put together a stunning itinerary. You might meet historical figures such as Buddha, Muhammad, or Moses. You could see what Cleopatra really looked like or attend Shakespeare's first production of *Hamlet*. You might position yourself on that grassy knoll in Dallas to see for yourself whether Oswald was the lone assassin. You might take in Jesus' Sermon on the Mount and even film it. You could enjoy an evening walk through the Hanging Gardens of Babylon. The possibilities are unlimited.

We seem free to move around in space at will, but in time we are like helpless rafters in a mighty stream, propelled into the future at the rate of one second per second. One wishes one could sometimes paddle ahead to investigate the shores of the future, or perhaps turn around and go against the current to visit the past. The hope that such freedom will one day be ours is bolstered when we observe that many feats formerly thought impossible have now been realized and are even taken for granted. When Wells wrote *The Time Machine* in 1895, many people thought that heavier-than-air flying machines were impossible. Eventually the Wright brothers proved the skeptics wrong. Then people said that we could never break the sound barrier. But Chuck Yeager ultimately proved that the seemingly impossible was possible. Flights to the Moon were confined to the realm of fantasy—until the Apollo program achieved it. Might time travel be similar?

Today the subject of time travel has jumped from the pages of science fiction to the pages of physics journals as physicists explore whether it might be allowed by physical laws and even if it holds the key to how the universe began. In Isaac Newton's universe time travel was inconceivable. But in Einstein's universe it has become a real possibility. Time travel to the future is already known to be permitted, and physicists are investigating time travel to the past as well. To appreciate what scientists are studying now, an excellent first step is to explore major time-travel themes in science fiction, where many ideas in this arena were first advanced.

THE TIME MACHINE AND TIME AS THE FOURTH DIMENSION

The idea of time travel gained prominence through Wells's wonderful novel. Most remarkable is his treatment of time as a fourth dimension, which anticipates Einstein's use of the concept ten years later.

The novel begins as the Time Traveler invites his friends to inspect his new invention—a time machine. He explains the idea to them:

"You know of course that a mathematical line, a line of thickness *nil,* has no real existence. . . . Neither has a mathematical plane. These things are mere abstractions."

"That's all right," said the Psychologist.

"Nor, having only length, breadth, and thickness, can a cube have a real existence."

"There I object," said Filby. "Of course a solid body may exist. All real things—"

". . . But wait a moment. Can an *instantaneous* cube exist?"

"Don't follow you," said Filby.

"Can a cube that does not last for any time at all, have a real existence?"

Filby became pensive. "Clearly," the Time Traveler proceeded, "any real body must have extension in *four* directions: it must have Length, Breadth, Thickness, and—Duration. . . . There are really four dimensions, three . . . of Space, and a fourth, Time. There is, however, a tendency to draw an unreal distinction between the former three dimensions and the latter because . . . our consciousness moves intermittently . . . along the latter from the beginning to the end of our lives."

The Time Traveler then shows his friends a small model of his invention—a metallic frame with ivory and quartz parts. One lever can propel it toward the future, and another can reverse the direction. He helps one of his friends push the future lever, and the model promptly disappears. Where did it go? It didn't move in space at all; it simply went to another time, the Time Traveler explains. His friends can't decide whether to believe him.

Next, the Time Traveler takes his friends to his home laboratory, to see his nearly complete, full-scale model. A week later he finishes the time machine, climbs aboard, and begins a remarkable journey to the future.

First he presses the future lever gently forward. Then he presses the one for stopping. He looks at his lab. Everything is the same. Then he notices the clock: "A moment before, as it seemed, it had stood at a minute or so past ten; now it was nearly half-past three!" He pushes the lever ahead again, and he can see his housekeeper flit across the room at high speed. Then he pushes the lever far forward. "The night came like the turning out of a light, and in another moment came tomorrow.... As I put on a pace, night followed day like the flapping of a black wing.... Presently, as I went on, still gaining velocity, the palpitation of night and day merged into one continuous grayness.... I saw huge buildings rise up faint and fair, and pass like dreams."

Eventually, the Time Traveler brings his vehicle to a stop. The machine's dials show that he has arrived in the year 802,701. What does he find? The human race has split into two species: one, brutish and mean, living below ground—the Morlocks; the other, childlike and gentle, living above ground—the Eloi. Among the aboveground dwellers he finds a lovely young woman named Weena, whom he befriends. He discovers, to his horror, that the troglodytes living below breed and harvest the gentle people above like cattle—to eat. To make matters worse, the Morlocks manage to steal his time machine. When he finds it, he jumps aboard, and to escape the Morlocks, he pushes the lever into the extreme forward position. By the time he is able to bring the machine under control, he has moved into the far future. Mammals have become extinct, and only some crablike creatures and butterflies remain on Earth. He explores as far as 30 million years into the future, where he discovers a dull red

Sun and lichen-like vegetation; the only animal life in evidence is a football-shaped creature with tentacles.

The Time Traveler then returns to his own time and to his friends. As proof of his experience in the future, he produces a couple of flowers Weena had given him, of a type unknown to his friends. After talking to his friends, the Time Traveler departs on his time machine and never returns. One friend muses about his fate. Where did he go? Did he return to the future or go instead to some prehistoric realm?

H. G. Wells's book was extraordinarily prescient in interpreting time as a fourth dimension. Einstein would use the idea in his 1905 theory of special relativity, which describes how time is measured differently by stationary and moving observers. Einstein's work, expanded by his mathematics professor Hermann Minkowski, shows that time can indeed be treated mathematically as a fourth dimension. Our universe is thus four-dimensional. By comparison, we say that the *surface* of Earth is two-dimensional because every point on Earth's surface can be specified by two coordinates—longitude and latitude. The universe, however, is four-dimensional. Locating an event in the universe requires four coordinates.

This example adapted from Russian physicist George Gamow further illustrates the point. If I want to invite you to a party, I must give you four coordinates. I may say the party will be at 43rd Street and 3rd Avenue on the 51st floor next New Year's Eve. The first three coordinates (43rd Street, 3rd Avenue, 51st floor) locate its position in space. Then I must tell you the time. The first two coordinates tell you where to go on the surface of the Earth, the third tells you how high to go, and the fourth tells you when to arrive. Four coordinates—four dimensions.

We may visualize our four-dimensional universe by using a three-dimensional model. Figure 1 shows such a model of the solar system. The two horizontal dimensions represent two

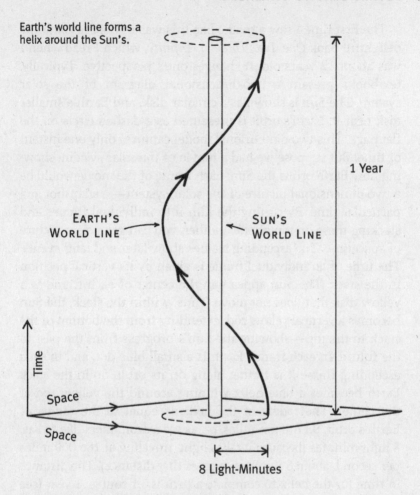

Earth's world line forms a helix around the Sun's.

EARTH'S WORLD LINE

SUN'S WORLD LINE

1 Year

Time

Space

Space

8 Light-Minutes

Figure 1. The Four-Dimensional Universe

dimensions of space (for simplicity, the third dimension of space is left out), and the vertical dimension represents the dimension of time. Up is toward the future; down is toward the past.

The first time I saw a model like this was in George Gamow's delightful book *One, Two, Three ... Infinity*, which I read when I was about 12 years old. It changes one's perspective. Typically, textbooks present a two-dimensional diagram of the solar system. The Sun is shown as a circular disk, and Earth a smaller disk near it. Earth's orbit is presented as a dashed circle on the flat page. This two-dimensional model captures only one instant of time. But suppose we had a movie of the solar system, showing how Earth orbits the Sun. Each frame of the movie would be a two-dimensional picture of the solar system—a snapshot at a particular time. By cutting the film into individual frames and stacking these on top of one another, you can get a clear picture of spacetime. The ascending frames show later and later events. The time of an individual frame is given by its vertical position in the stack. The Sun appears in the center of each frame as a yellow disk that does not move. Thus, within the stack, the Sun becomes a vertical yellow rod, extending from the bottom of the stack to the top—showing the Sun's progress from the past to the future. In each frame, Earth is a small blue dot, and in each ascending frame it is farther along on its orbit. So in the stack Earth becomes a blue helix winding around the yellow rod at the center. The radius of the helix is equal to the radius of Earth's orbit, 93 million miles, or, as we astronomers like to say, 8 light-minutes (because it takes light, traveling at 186,000 miles per second, about 8 minutes to cross that distance). The distance in time for the helix to complete a turn is, of course, 1 year (see Figure 1). This helix is Earth's *world line*, its path through spacetime. If we were to think four-dimensionally, we would see that Earth is not just a sphere—it is really a helix, a long piece of spaghetti spiraling around the Sun's world line through time.

As the Time Traveler said, all real objects have four dimensions—width, breadth, height, and duration. Real objects have an extension in time. Your dimensions are perhaps 6 feet tall, 1 foot thick, 2 feet wide, and 80 years in duration. You have a world line too. Your world line starts with your birth, snakes through space and forward in time, threading through all the events of your life, and ends at your death.

A time traveler who visits the past is just someone whose world line somehow loops back in time, where it could even intersect itself. This would allow the time traveler to shake hands with himself. The older man could meet up with his younger self and say, "Hi! I'm your future self! I've traveled back in time to say hello!" (see Figure 2). The surprised younger man would reply, "Really?" He would then continue his life, becoming old and eventually looping back to that same event—where he would recognize his younger self, shake hands, and say, "Hi! I'm your future self! I've traveled back in time to say hello!"

BACK TO THE FUTURE AND THE GRANDMOTHER PARADOX

But what if, as an older man, the time traveler refuses to say hello and instead simply kills his younger self? Time travel to the past suggests such a paradox. When I do television interviews about time travel, the first question I am always asked is this: "what if you went back in time and killed your grandmother before she gave birth to your mother?" The problem is obvious: if you kill your grandmother, then your mother would have never been born, and you would never have been born; if you were never born, you could never go back in time, and so you could not kill your grandmother. This conundrum, known as the Grandmother Paradox, is often thought sufficiently potent to rule out time travel to the past.

A famous example from science-fiction stories that have explored this idea is the 1985 movie *Back to the Future*. The hero, played by Michael J. Fox, goes back in time to 1955 and accidentally interferes with the courtship of his parents. This

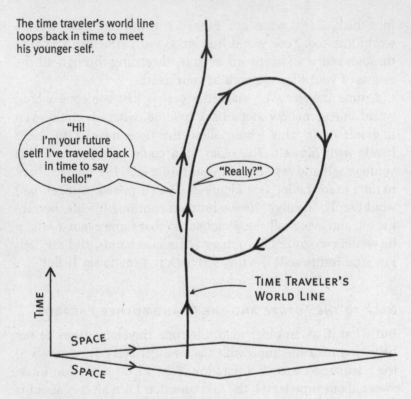

The time traveler's world line loops back in time to meet his younger self.

"Hi! I'm your future self! I've traveled back in time to say hello!"

"Really?"

TIME TRAVELER'S WORLD LINE

TIME

SPACE

SPACE

Figure 2. Meeting a Younger Self in the Past

creates a problem: if his parents don't fall in love, he will never be born, so his own existence is imperiled. He realizes he must act to ensure that his parents fall in love. Things don't go well at first—his mother begins to fall in love with *him*, the mysterious stranger, instead of his father. (Freud, take note.) To bring his parents together, he hatches an elaborate plan. He realizes it is failing when the images of himself and his brother and sister vanish from the family picture he carries in his wallet—a bad sign. Later he sees his own hand fading away. He can look right through it. He is disappearing. He begins to feel faint. Because he has interrupted his parents' romance, he is slipping out of existence. Later, when his plan finally succeeds and his parents are united, he suddenly feels better and his hand returns to normal. He looks in his wallet; the pictures of himself and his brother and sister have reappeared.

A hand can fade in a fictional story, but in the physical realm, atoms just don't dematerialize that way. Besides, according to the parameters of the story, the boy is dematerializing because, as a time traveler, he prevented his parents from falling in love, thereby circumventing his own birth. But if he was never born, his entire world line, from the point of his birth to his adventures as time traveler, should vanish, leaving no one to interfere with his parents—so his birth would have happened after all. Clearly, this fictional story has not resolved the Grandmother Paradox. Physically possible solutions to such time-travel paradoxes exist, but physicists are divided on which of two approaches is correct.

TIMESCAPE AND THE MANY-WORLDS THEORY

First, the radical alternative. It involves quantum mechanics, that field of physics developed in the early twentieth century to explain the behavior of atoms and molecules. Quantum mechanics shows how particles have a wave nature, and waves have a particle nature. A key feature is Heisenberg's uncertainty principle, which tells us that we cannot establish a particle's position and velocity with arbitrary accuracy. Such quantum fuzziness, although usually negligible in the macroscopic world, is important on atomic scales. Quantum mechanics explains how atoms emit or absorb light at specific wavelengths when electrons jump from one energy level to another. The wave nature of particles leads to unusual effects such as quantum tunneling, in which a helium nucleus may suddenly jump out of a uranium nucleus, causing its radioactive decay. Solving quantum wave equations allows you to predict the probability of finding a particle at various places. This in turn leads, in one interpretation, to the many-worlds theory of quantum mechanics, which posits different parallel worlds where the particle is detected at those various places. Many physicists think this interpretation is an unnecessary addition to the theory, but a number of physicists working on the frontiers of our understanding of quantum theory do take this many-worlds interpretation and its refinements and extensions seriously.

In this picture, the universe contains not one single world history but many in parallel. Experiencing one world history, as we do, is like riding a train down a track from the past to the future. As passengers on the train, we see events go by like stations along the track—there goes the Roman Empire, there goes World War II, and look, people are landing on the Moon. But the universe might be like a giant switching yard, with many such railroad tracks interlaced. Next to our track is one on which World War II never happened. A train is constantly encountering switches at which it may take either of two lines. Before World War II, there may have been a day when Hitler could have been killed, diverting the train onto a track on which World War II did not occur. According to the many-worlds theory of quantum mechanics, a branch in the tracks occurs every time an observation is recorded or a decision is made. It doesn't have to be a human observation or decision; even an electron in an atom making a change from one energy level to another could cause a branching of the track.

In this scenario, in Oxford University physicist David Deutsch's view, a time traveler may go back in time and kill his grandmother when she was a young girl. That will cause the universe to branch onto a different track that contains a time traveler and a dead grandmother. The universe in which the grandmother lived and gave birth to the mother who in turn gave birth to the time traveler—the universe we remember seeing—still exists. For it is from that universe (that track) that the time traveler came. The time traveler just moves to a different universe, where he will participate in a changed history.

These ideas are illustrated well by Gregory Benford's 1980 Nebula Award–winning sci-fi novel *Timescape*. The story is set in 1998; its hero uses a beam of tachyons—hypothetical particles that move faster than light—to send a signal to 1963, warning scientists of an ecological disaster that will engulf the world of 1998.

This novel came to my attention because a 1974 paper of mine appears in it. The hero reads my paper during an airplane trip in 1998, which gives him an important clue for making his tachyon transmitter. As Benford puts it, "He rummaged through his briefcase for the paper by Gott that Cathy had given him. Here: *A Time-Symmetric, Matter and Anti-Matter Tachyon Cosmology*. Quite a piece of territory to bite off, indeed. But Gott's solutions were there, luminous on the page." (Would that all my research papers shone so brightly!)

The warning is received in the fall of 1963, and the scientists begin to act on it. They know about the many-worlds theory of quantum mechanics, and their publication of the warnings about the ecological disaster helps avert it, by sending the universe onto a track on which the disaster is avoided. Incidentally, in that parallel universe, President Kennedy is only wounded in Dallas, rather than killed.

Of course, this is just a work of fiction. Or is it? Maybe there is some parallel universe in which everything happened just as the book describes it.

Why would some people believe that an infinite number of parallel universes exist, playing out all possible world histories, despite the fact that we ourselves actually observe only one world history? The celebrated California Institute of Technology (Caltech) physicist Richard Feynman showed that, in general, if one wished to calculate the probability of a certain outcome, one had to consider all possible world histories that could lead up to it. So perhaps all the world histories are real.

To someone hoping to find a time machine in order to go to the past to save a lost loved one, the most comforting thing I can say is that, as far as we understand today, this can only be accomplished if the many-worlds theory of quantum mechanics is true. And if that is true, then there is *already* a parallel universe in which your loved one is okay now. That's because all the possible universes exist. Unfortunately, you are just in the wrong one.

BILL AND TED'S EXCELLENT ADVENTURE AND SELF-CONSISTENCY

Now for the more conservative approach to the Grandmother Paradox: time travelers don't change the past because they were always part of it. The universe we observe is four-dimensional, with world lines snaking through it. If some of these world lines can bend back and cross through the same event twice, then so be it. The time traveler can then shake hands with an earlier version of himself. The solution has to be self-consistent, however. This *principle of self-consistency* has been advanced by physicists Igor Novikov of the University of Copenhagen, Kip Thorne of Caltech, and their collaborators. In this case, the time traveler may have tea with his grandmother while she is a young girl, but he can't kill her—or he would not be born, and we already know he was. If you witness a previous event, it must play out just as before. Think of rewatching the classic movie *Casablanca*. You know how it's going to turn out. No matter how many times you see it, Ingrid Bergman always gets on that plane. The time traveler's view of a scene would be similar. She might know from studying history how it is going to turn out, but she would be unable to change it. If she went back in time and booked passage on the *Titanic*, she would not be able to convince the captain that the icebergs were dangerous. Why? Because we know already what happened, and it cannot be changed. If any time travelers were aboard, they certainly failed to get the captain to stop. And the names of those time travelers would have to be on the list of passengers you can read today.

Self-consistency seems contrary to the common sense notion of free will. Though we seem to experience free will, to be able to do what we please, the time traveler seems constrained. This seems to rob the time traveler of an essential human ability. But consider this. Free will never did allow one to do something logically impossible—an important point made by Princeton philosopher David Lewis in analyzing time-travel paradoxes. I might wish right now to instantly become a tomato larger than the whole universe, but no matter how hard I try, I cannot do it. Killing your grandmother as a young girl during a time-travel expedition may be a similarly impossible task. If you think of the universe as one four-dimensional entity with world lines winding through it like so many garden hoses, it is clear why. This four-dimensional entity does not change—it is like an intricate, fixed sculpture. If you want to know what it is like to experience living in that universe, you must look along the world line of a particular person from beginning to end.

Many science-fiction time-travel stories have explored the concept of a self-consistent world history. The charming 1989 movie *Bill and Ted's Excellent Adventure* has a lot of fun with the idea. Bill and Ted are two high school boys hoping to form a rock band. Unfortunately, they are failing history, and if they don't pass, Ted will be sent to military school in Alaska, splitting up their band. Their only hope is to get an A$^+$ in their upcoming history presentation, but they are clueless about what to do.

Then a time traveler from the year 2688 (played by George Carlin) arrives. Apparently, the music and lyrics produced by Bill and Ted's rock band form the foundation of a great future civilization. These lyrics include sayings like "Be excellent to each other" and "Party on, dudes!" Thus, the time traveler has come to help them on their history project so their rock band can indeed be formed. He provides them with a time machine that looks exactly like a phone booth. Just after meeting the time traveler from the future, Bill and Ted encounter slightly older versions of themselves who have returned to the present. Now the younger Bill and Ted are convinced that they're on their way to a history project that will make history and keep their band together. They decide to go to the past and pick up some historical figures to bring to their history assembly, making their project exciting enough to garner an A$^+$.

As we follow Bill and Ted's adventure, we see this scene played out again, this time when Bill and Ted are their older selves. The scene unfolds exactly as it did before. So far, so good. No time-travel paradoxes.

Bill and Ted use the time machine to round up a number of historical figures: Napoleon, Billy the Kid, Freud, Beethoven, Socrates, Joan of Arc, Lincoln, and Genghis Khan. They bring them to twentieth-century California, and chaos ensues. The historical figures get into trouble in the San Dimas Mall. Beethoven draws a crowd by playing the electric organ in the music store, Joan of Arc gets arrested after taking over an aerobics class, and Genghis Khan trashes a sporting goods store while testing a baseball bat as a weapon. Eventually, the historical figures land in jail. As these events unfold, time is running out, leaving only a few minutes until Bill and Ted's history presentation is due.

Luckily, Ted's father is the sheriff, and Ted remembers his father had keys to the jail a couple of days ago, before he lost them. Bill suggests using the time machine to go back and get them, but, unfortunately, there is not enough time to get to their time machine before the history assembly starts. Then Ted has a great idea. Why not just make sure, *after* the assembly, to go back in time and steal the keys? Then they could leave them hidden nearby, say, behind a particular sign, Bill suggests. Bill reaches behind the sign. There they are! They take the keys, break Genghis Khan and the others out of jail—leaving the keys with Ted's astonished father—and arrive at the school auditorium with their historical figures, just in time to make their presentation before a cheering audience. They, of course, get an A+, and the emergence of a splendid, rock-inspired future civilization is ensured. The boys must now remember to go back in time, find the keys, and hide them behind the sign.

Did Bill and Ted exercise free will? Well, it certainly appeared so to them. When, in the course of their adventures, they arrived to meet their younger selves, they wondered about the upcoming conversation. They didn't remember what they had said, so they proceeded with the meeting—which, of course, went exactly as before. They were always doing what they wanted to do, but their actions appear to have been fated. Once they found those keys behind the sign, they had to go back in time, steal the keys, and plant them there, didn't they?

Though they can sometimes be complicated, self-consistent histories such as this one are possible, and a number of stories about time travel to the past have illustrated them.

Self-consistency is the conservative possibility: you can visit the past, but you can't change it. I personally find this view the most attractive. One reason is that arriving at self-consistent solutions—in fact, numerous ones—always seems possible from a given set of starting conditions, as suggested by

Thorne, Novikov, and their collaborators in an elaborate series of thought experiments involving billiard balls going back in time. They tried to produce situations where a time-traveling billiard ball would collide with its earlier self, deflecting its trajectory so it couldn't enter the time machine in the first place. But they could always find a self-consistent solution where the collision was only a light tap that didn't stop the ball from entering the time machine, but sent it on a path that made it nearly miss its earlier self and only administer that light tap, instead of a heavy blow. No matter how hard the physicists tried to produce paradoxes, they always found it possible to find self-consistent solutions from a given start. Following Thorne and his colleagues, those who hold the conservative view believe that *even* in the many-worlds picture, one would still expect the principle of self-consistency to be upheld— each track in the switching yard must be self-consistent. However, many self-consistent ways of playing out an event may exist in parallel, some involving time travelers. In each parallel universe, different things happen. In some, for example, the time traveler has tea with her young grandmother, whereas in others she sips lemonade. But each track is self-consistent, and in each, the time traveler never kills the grandmother. Each time traveler finds it impossible to change the past she remembers.

SOMEWHERE IN TIME AND THE IDEA OF JINN

Even time-travel stories based on the concept of self-consistency can have some curious features, however. Generally we think of a person's or particle's world line as snaking through time, with a beginning and an end. But in time travel, it is possible for a particle to have a world line that looks like a hula hoop—a circle with no ends. Such particles are called *jinn* by

Igor Novikov. Like Aladdin's genie (from the Arabic *jinni*, from which Novikov derives the term), they seem to arrive by wizardry. The watch in the 1980 movie *Somewhere in Time*, starring Christopher Reeve and Jane Seymour, is an example.

The story begins in 1972. Christopher Reeve is a young playwright being congratulated after the opening night of his play. An old woman from the audience approaches him and gives him a gold watch: "Come back to me," she says enigmatically before leaving. Eight years later, in 1980, he takes a vacation at the Grand Hotel on Mackinac Island, Michigan. In the hotel he sees an old framed photograph of a beautiful young woman. He falls instantly in love with the woman in the picture. He asks the elderly bell captain who she was. The bell captain tells him that she was Elise McKenna, a famous actress who performed at the hotel in 1912. The playwright tries to find out about this woman. On a trip to the library, he finds a magazine article containing the last picture ever taken of her—why, it's the mysterious old woman who gave him the gold watch! Now he is really hooked. He visits the author of a book on distinguished actresses and learns that Elise McKenna died on the night she gave the playwright the watch. He also discovers that she especially cherished a book on time travel.

The playwright then seeks out the professor who wrote that book. The professor's theory of time travel involves self-hypnosis. He hypothesizes, for example, that if you go to an old hotel, dress up in a period costume, use your imagination hard enough, and chant continuously the time you wish to visit, you may be transported to the past. The professor had tried it once and felt that he had been transported back, but the impression lasted for only a moment, so he could never prove it.

Now, eager to test the technique himself, the playwright returns to the hotel and reviews the old guest books to pinpoint the exact day in 1912 that the young Miss McKenna checked in. He finds the very page she signed. In the same book, he finds his own signature! He was there. With this encouragement, he dons a suit from that period and takes along that gold watch. He lies in bed in the hotel—after stowing in the closet every modern article in the room that might disrupt his concentration on the past. Over and over he chants the day in 1912 he wants to visit—and drifts off to sleep. He wakes up—you guessed it—surrounded by the ornate décor of a 1912 hotel room.

Never mind how this is accomplished physically. The young man goes to register at the desk at the exact time, 9:18 A.M., that he had read in the hotel guest book. He wants to make his signature in the guest book correct because he is afraid that if he does it wrong, he will break the spell and wake up back in 1980. He wants to fulfill the past, not change it. He meets Miss McKenna, who is performing in a play at the hotel, and, not surprisingly, they soon fall deeply in love. In fact, he is there when she has her photograph taken; she looks up at him lovingly at just the moment when the picture is snapped. After a night of lovemaking, they plan their future together. She picks up the gold watch to check the time. She teases him about his suit, saying it's at least 15 years old. He playfully objects, bragging that it has a great pocket for coins. He pulls out a penny and notices it bears the date 1979. He has made a mistake! A modern coin has somehow slipped into the pocket. He reaches out to her, but she and the whole room fade quickly into the distance, and he finds himself back in the hotel in 1980. (Oh, dear.) He tries desperately chanting the appropriate date in 1912, but it doesn't work. He can't get back anymore. He pines away and soon dies of a broken heart—whereupon he is greeted, of course, by a young Miss McKenna, and they are enveloped in a white light. Music up, credits roll.

Although the time-travel mechanism leaves implausible gaps,

the story otherwise takes great care to be self-consistent. There are no paradoxes. Christopher Reeve's character does not alter the past at all—he fulfills it. He participates in the past, making Miss McKenna fall in love with him and bringing her the watch that she will later, as an old woman, give to him.

But where did the watch come from? This watch is a jinni— elderly Miss McKenna gives it to the young playwright, who takes it back in time to deliver it to her as a young woman. She keeps it all her life until it is time to return it to him. So who made the watch? No one. The watch never went anywhere near a watch factory. Its world line is circular. Novikov has noted that in the case of a macroscopic jinni like this the outside world must always expend energy to repair any wear-and-tear (entropy) it has accumulated so it can be returned exactly to its original condition as it completes its loop.

Permissible in theory, macroscopic jinn are improbable. The whole story in *Somewhere in Time* could have taken place without the watch. The watch seems particularly unlikely since it appears to keep good time. One could have imagined finding a nonworking watch or perhaps a paper clip that passes back and forth between the couple. How lucky to encounter a watch that works! According to quantum mechanics, if one has enough energy, one can always make a macroscopic object spontaneously appear (along with associated antiparticles, which have equal mass but opposite electric charge)—it's just extremely unlikely. Similarly with jinn, it would be more improbable to find a watch than a paper clip and more improbable to find a paper clip than an electron. The more massive and more complex the macroscopic jinni, the rarer it will be.

Novikov has pointed out that even information traveling in a closed loop can constitute a jinni, even though no actual particles have circular world lines. Suppose I went back in time to 1905 and told Einstein all about special relativity. Then Einstein could publish it in his paper in 1905. But I learned about special relativity by reading about Einstein's paper later. Such a scenario is possible, but highly unlikely. Jinn remain intriguing nevertheless.

"ALL YOU ZOMBIES—" AND HUMAN SELF-CREATION THROUGH TIME TRAVEL

Even more intriguing is one of the most remarkable time-travel stories ever written, "All You Zombies—" (1959), by science-fiction master Robert Heinlein. A 25-year-old man is in a bar lamenting his fate; curiously, he calls himself the "Unmarried Mother." He tells the bartender his story. This man has had it rough. He had been born a girl and raised in an orphanage. As a young woman, she had had sex with a man who then abandoned her. She became pregnant and decided to keep the baby. When it came time to give birth, she had a cesarean section. The baby was born—it was a girl. During the operation, the doctor noticed that the woman had, hidden inside her body, male as well as female organs. With some reconstructive surgery, the doctor transformed her into a man without her consent. This is why the man refers to himself as the "Unmarried Mother." Moreover, the child was soon kidnapped from the hospital by a stranger.

The bartender interrupts the young man's story: "The matron at your orphanage was Mrs. Fetherbridge—right? . . . Your name as a girl was Jane—right? And you didn't tell me any of this—right?" The bartender asks the Unmarried Mother whether he wants to find the man who had gotten "him" pregnant. He does. Then the bartender ushers the unfortunate young man to the rear of the bar to a time machine. They go back in time 7 years and 9 months, where the bartender drops the man off. The bartender then goes forward in time 9 months, just in time to abduct a baby named Jane. He next takes baby Jane back 18 years earlier in time and puts her on the steps of an orphanage. After that he returns to the young man, who has just impregnated a young woman named Jane. The bartender then takes the young man to the future to learn the trade of bartending. At the end, the bartender considers the whole affair, and looks down at his old cesarean scar: "I know where I come from—*but where did all you Zombies come from?*" he muses.

The bartender, who is Jane, has gone back in time to become both his own mother and father. His world line is indeed complex. He starts as baby Jane, is taken back in time by a bartender, grows up in an orphanage, has sex with a man, gives birth to a girl named Jane, changes sex, goes to a bar to lament his fate, takes a trip back in time with a bartender, has sex with a woman named Jane, and is picked up by the bartender and taken to the future, where he becomes a bartender who then travels back in time to engineer the whole thing. It is a self-consistent story, both bizarre and wonderful.

Carried to the species level by Ben Bova in his 1984 novel *Orion*, time travel allows humans from the future to go back in time and start the human race. Thus, in the story, the human race creates itself. In similar fashion, I later consider how time travel in general relativity may allow the universe to be its own mother.

CONTACT AND THE CONCEPT OF WORMHOLES

Sometimes science fiction leads directly to a scientific investigation. In 1985 Carl Sagan was writing a science-fiction novel called *Contact* (later made into a movie starring Jodie Foster). Sagan wanted his heroine to fall into a small black hole on Earth and pop out of another black hole far away in space.

He asked his friend Caltech professor Kip Thorne to check whether the fictional account he was writing violated any physical laws. Thorne said that what Sagan really wanted was a wormhole—a spacetime tunnel—connecting the two locations. Thorne thus became interested in the physics of wormholes and, with his colleagues, showed how they might be used to travel to the past.

Sagan wished to show, in dramatic fashion, the profound consequences of contact with an extraterrestrial civilization. In the movie, Jodie Foster plays a SETI (Search for Extraterrestrial Intelligence) scientist who hears a radio signal while monitoring the star Vega. She notifies a colleague in Australia who finds he can simultaneously observe it with his radio telescope. After the confirmation, her assistant asks, "Who do we call now?" "Everybody," Foster replies. Soon, everyone from CNN to the president of the United States is involved. It becomes clear that the signal is actually a TV transmission, so Foster puts it up on a monitor. It's a picture of Hitler addressing a Nazi rally. Nazis on Vega? No, the Vegans are just sending back a TV signal they had received from Earth, part of an early broadcast sent out in 1936. Vega is 26 light-years away, so it took that TV signal 26 years, traveling at the speed of light, to reach Vega. When the Vegans received our signal, it alerted them to the presence of intelligent life on Earth. (What a bad first impression we must have made.) The Vegans had apparently figured that we would have an easy time interpreting our own signal, making it an ideal calling card with which to announce their own presence. So the Vegans just duplicated our signal and sent it back to us. That reply took another 26 years to arrive back on Earth in 1988. A second set of pictures interleaved with the frames of the TV broadcast reveals a complicated set of blueprints. They appear to be instructions for building some kind of spaceship—a sphere with a place for a person inside.

Should this spaceship be built? A heated debate follows: it might not be a spaceship at all but a bomb to blow up Earth. Finally, the extraterrestrials are presumed benevolent, so the spaceship is constructed according to the plans. Jodie Foster gets to be the astronaut. Once she is inside the sphere, the door closes, and—Bam! It creates a wormhole connecting directly to a location in the Vegan star system. The spaceship falls through the wormhole and emerges near Vega. Foster sees the Vegan system, then is whisked off via another wormhole to an encounter with an extraterrestrial, who assumes the likeness of her father. Finally, she returns via the wormholes to Earth. Surprisingly, she learns that she has returned at exactly the same time she left. As she exits the sphere, the launch team asks why it didn't work. According to Foster, her trip had taken 18 hours, but according to the people outside, her trip took zero time— as far as they could tell, the ship never left. Thus, many pundits refuse to believe her account. At the end of the movie, however, we find out that the president's national security adviser has noticed something: although Foster's video camera failed to record pictures that would verify her story, it did record exactly 18 hours of static. So he knows that she really went somewhere, but we are left with the idea that the adviser will keep this secret.

When Sagan came up with his basic plot, he asked Kip Thorne whether wormholes could really allow the plot in principle—even if it required superadvanced technology. Wormholes connected with black holes had already been discussed in the scientific literature. The trouble was that the wormhole pinched off so fast that there was never enough time for a spaceship to traverse it from one end to the other without being crushed. Kip Thorne and his colleagues then thought up a physically logical way to prop the wormhole open with exotic matter (in lay terms, stuff that weighs less than nothing) to allow travel through it without the risk of being crushed. Then they made a fascinating discovery—a way to manipulate the two ends of the wormhole so that Jodie Foster's character could not only return at the exact instant she started, but even earlier. Here was a time machine allowing one to visit the past. Thorne and his colleagues published their results in the eminent journal *Physical Review Letters* in 1988, sparking a new interest in time travel.

STAR TREK AND THE CONCEPT OF WARPDRIVE

Another example of science fiction stimulating scientific investigation comes from *Star Trek*, which has featured so many time-travel stories that they are hard to count. *Star Trek* is set in the twenty-third century and chronicles the adventures of the crew of the starship *Enterprise*. Originally a TV series, it spawned a number of successful movies and several spinoff TV series, becoming enshrined as a cultural classic. Gene Roddenberry, who created the series, wanted to tell a story of interstellar travel in which the *Enterprise* would visit a different star system each week and return to Star Fleet Headquarters, to report the results of their explorations, all within a 5-year period. To allow the *Enterprise* to move at a speed far faster than that of light, he used the idea of warpdrive. Somehow, the space around the ship would warp, or bend, allowing the ship to zoom between stars in short order. At the time when the series was created (the mid-1960s), most physicists would have scoffed at the idea as pure fantasy. Then Miguel Alcubierre, a Mexican physicist, decided to see if the idea could work according to the rules of Einstein's gravitational theory. It could, but it required the presence of some exotic matter (as do Thorne's wormholes). Alcubierre's solution, published in 1994, did not involve time travel to the past, but he speculated that, if one were clever enough, a warpdrive might be used to visit the past. Two years later, a paper by physicist Allen E. Everett showed how to accomplish this by applying the warpdrive twice in succession.

Interestingly, the writers of *Star Trek* always seemed to know instinctively that the warpdrive could be used to visit the past, and they incorporated this idea into many episodes. One of the best stories of this kind plays out in the movie *Star Trek IV: The Voyage Home*. A crisis arises in the twenty-third century when a giant extraterrestrial spaceship arrives and starts warming up a giant death ray to destroy Earth. The ship is sending out a signal: a song of humpback whales. The extraterrestrials make it clear (to listening humans) that if they do not receive a suitable reply from a humpback whale, they are going to destroy Earth. Unfortunately, humpback whales had become extinct by the twenty-third century, so there aren't any left to answer the signal. The solution: use the warpdrive to somehow slingshot back to the twentieth century when humpback whales existed, retrieve a couple of whales, and bring them back to the twenty-third century just in time to sing an answer back to the extraterrestrials, so the monster spaceship with its death ray can go nicely away.

As you can see, science fiction often gets scientists thinking.

CHESS AND THE LAWS OF PHYSICS

Why are physicists like me interested in time travel? It's not because we are hoping to patent a time machine in the near future. Rather, it's because we want to test the boundaries of the laws of physics. The paradoxes associated with time travel pose a challenge. Such paradoxes are often a clue that some interesting physics is waiting to be discovered.

Einstein addressed some existing paradoxes in creating his theory of special relativity. Physicist Albert Michelson and chemist Edward Morley had done a beautiful experiment in 1887, showing that the velocity of light was exactly the same, regardless of the direction in which it was traveling in their lab. But this phenomenon should occur only if Earth was stationary, and scientists knew that the Earth circles the Sun. This presented a paradox. Einstein solved it by developing his theory of special relativity, which, as we shall see, overthrew Isaac Newton's conception of space and time. The atomic bomb proved in dramatic fashion that the theory works, confirming its key equation, $E = mc^2$, by showing that a little bit of mass could be converted into an enormous amount of energy.

Quantum mechanics, the field that Einstein himself had qualms about but that physicists have since embraced, has its own paradoxes. Yet quantum mechanics works. It can predict the probabilities of obtaining different outcomes of an experiment. Naturally, if you add up the predicted probabilities of all possible outcomes of a given experiment, you should automatically get a total of 100 percent. But David Boulware of the University of Washington, working on a time-travel solution I discovered, later showed that jinn particles prevent the probabilities from adding up to 100 percent. My former student Jonathan Simon and his colleagues addressed this paradox by showing that one could simply multiply the quantum probabilities by a correction factor so that they again add up to 100 percent. This investigation led Simon and his colleagues to favor Feynman's many-histories approach to quantum mechanics because it gave unique answers. Stephen Hawking thought of a different way around the problem. If some ways of doing quantum mechanics are flexible enough to work even across regions of time travel, we might well be tempted to regard them as more fundamental. This is why time-travel research is particularly interesting—it may lead to new physics.

Richard Feynman once noted that discovering the laws of

physics is like trying to learn the laws of chess merely by observing chess games. You notice that bishops stay on the same color squares; you write this down as a law of chess. Later, you come up with a better law—bishops move diagonally. And, since diagonal squares are always colored the same, this explains why bishops always stay on the same color. This law is an improvement—it is simpler, and yet explains more. In physics, discovering Einstein's theory of gravity after knowing Newton's theory of gravity is a similar type of discovery. As another example, noticing that pieces don't change their identity in a chess game is similar to discovering the law of mass-and-energy conservation.

Eventually, say, you see a chess game in which a pawn reaches the other end of the board and is promoted to become a queen. You say, "Wait, that violates the laws of chess. Pieces can't just change their identity." Of course, it does not violate the laws of chess; you just had never seen a game pushed to that extreme before. In time-travel research we are exploring extreme situations in which space and time are warped in unfamiliar ways. That these time-travel solutions may violate "common sense" makes them intriguing.

In the same way, quantum mechanics and special relativity violate common sense beliefs and yet have been confirmed by many experiments. Quantum mechanics violates our expectations of everyday life because we are used to dealing with objects that are so large and massive that quantum mechanical effects are minimal. You have never seen your car "tunnel" out of a closed garage. You never find your car just suddenly sitting out on the lawn. If someone told you that such a thing could occur (with a small but finite probability), you might—before the twentieth century—have argued that the laws of physics do not allow such effects. And yet this has been shown to be true on the subatomic scale; a helium nucleus may tunnel out of a uranium nucleus in precisely this fashion, as shown by George Gamow. Quantum tunneling seems strange because in the ordinary world of large massive objects, quantum effects are hardly ever important. Gamow wrote a popular book to emphasize this point, called *Mr. Tompkins in Wonderland* (now reprinted with the wonderfully quirky name *Mr. Tompkins in Paperback*). It shows how the world would look to us if the velocity of light were only 10 miles per hour and if quantum effects were important on everyday scales. Hunters would have to aim at fuzzy tigers that could not be located exactly. And you would always be losing your car when it tunneled unexpectedly outside your garage (not to mention those car keys we lose so easily). If you were used to seeing such things, they might not seem strange.

Time travel seems strange because we are not accustomed to seeing time travelers. But if we saw them every day, we might not be surprised to meet a man who was his own mother and father. Learning about whether time travel could occur in principle may give us new insights into how the universe works— and even how it got here.

2 TIME TRAVEL TO THE FUTURE

> A journey of a thousand miles must begin with
> a single step. — LAO-TZU

TIME TRAVEL TO THE FUTURE IS POSSIBLE

Do you want to visit Earth 1,000 years from now? Einstein showed how to do it. All you have to do is get in a spaceship, go to a star a bit less than 500 light-years away, and return, traveling both ways at 99.995 percent of the speed of light. When you come back, Earth will be 1,000 years older, but you will be only 10 years older. Such speed is possible—in our largest particle accelerators we bring protons to speeds higher than this (the best so far has been 99.999946 percent of the speed of light, at Fermilab).

We've already seen how naysayers of the past were wrong about heavier-than-air flying machines and breaking the sound barrier. They should have known better. As Leonardo da Vinci understood, birds fly despite being heavier than air, so building heavier-than-air flying machines should be possible. Likewise, when you crack a whip, the "crack" you hear is the little sonic boom created when the whip's tiny end breaks the sound barrier. Granted, the tip of the whip is very small compared with the size of an airplane, but the crack proves the possibility of exceeding the speed of sound. NASA, take note: if we can accelerate protons to greater than 99.995 percent of the speed of light, we could also send off an astronaut at the same speed. It's just a matter of cost. Protons don't weigh much, so accelerating them to high speed is relatively inexpensive. But since a human being weighs about 40 octillion times as much as a proton, in terms of energy alone, sending a person would be a great deal more expensive than sending a proton.

Of course, travel at nearly the speed of light would have to be planned to avoid too much wear and tear on the human body. For example, if we wanted to avoid extreme accelerations, we could simply limit the astronaut's acceleration to 1g (the acceleration of gravity on Earth). With this acceleration, as the rocket picked up speed, the astronaut's feet would be pressed against the floor, making her feel as though she weighed just as much as she does on Earth, thus ensuring that the trip would be quite comfortable. The astronaut would age 6 years and 3 weeks while accelerating up to a speed of 99.9992 percent of the speed of light, at which point she would be 250 light-years away from Earth. She would then turn her rocket around and fire it, and that reverse thrust would slow her down. She would age another 6 years and 3 weeks while slowing back down to zero velocity and continuing outward for another 250 light-years. She would thus arrive at a star 500 light-years away, having aged 12 years and 6 weeks. She would repeat this process on the return trip, aging another 12 years and 6 weeks. Earth would be 1,000 years older when she returned, but she would have aged fewer than 25 years during the trip.

Here's one scenario for how such a trip might be accomplished. The astronaut's capsule would weigh, say, 50 tons, and her multistage rocket, loaded with even the most efficient matter-antimatter fuel, would have to weigh more than 4,000 times as much as the Saturn V rocket. Here's how matter-antimatter fuel works. For every particle of matter (proton, neutron, or electron) there exists a corresponding particle of antimatter (antiproton, antineutron, or positron). Bring a particle of matter together with a particle of antimatter, and they will annihilate each other, producing pure energy usually in the form of gamma-ray photons. On the back of the rocket would be a large mirror—a light sail. To launch the capsule from Earth, a giant laser positioned in the solar system would fire at this mirror, accelerating the rocket's travel away from the solar system for the first quarter of the journey. The rocket would then be racing away from Earth at 99.9992 percent of the speed of light. The astronaut would then turn her rocket around and, in its engines, matter and antimatter would annihilate each other to produce gamma rays that exit out the back, slowing the rocket to a halt after another 250 light-years. Then the matter-antimatter engines would fire again, accelerating the rocket back up to speed for the return journey. Finally, the astronaut would pull out another mirror, and the laser stationed in the solar system would fire at it, efficiently slowing down the rocket for its return to Earth. This project would require space-based lasers vastly more powerful than those available currently. Also, at present we can make antimatter one atom at a time; we would have to be able to make it and store it safely in bulk. We would have to develop technology for cooling the engines to prevent melting. The ship would need to be shielded from interstellar atoms and light waves it would run into. There would be many serious engineering problems. It wouldn't be easy, but it is scientifically possible for a person to indeed visit the future.

EINSTEIN'S STUDY OF TIME AND THE SPEED OF LIGHT

Einstein's prediction that moving objects age slowly has been confirmed by experiment many times. One of the first demonstrations involved the decays of rapidly moving muons. Discovered in 1937, muons are elementary particles weighing about one tenth as much as protons. Muons are unstable—they decay into lighter elementary particles. If you observe a bunch of muons in the lab, you will find that only half are left after about two millionths of a second. But muons created in cosmic ray showers in the upper atmosphere and traveling at nearly the speed of light were not observed to decay as rapidly on their way to Earth's surface as those in the lab did, in accordance with Einstein's predictions. In 1971, physicists Joe Hafele and Richard Keating demonstrated Einstein's slowing of time in moving objects by taking very accurate atomic clocks on an airplane trip east around the world, a journey in which the plane's velocity adds to that of Earth's rotation. The physicists observed that the clocks were slightly slow—by 59 nanoseconds—relative to clocks on the ground when they returned, an observation in exact agreement with Einstein's predictions. (Because of Earth's rotation, the ground is also moving, but not as fast. Clocks on the ground are slowed less than those on the plane.)

Einstein began thinking about the nature of time and its relation to the speed of light while still a teenager. He imagined that if, starting at noon, he flew away from the big clock in the town square at the speed of light and looked back at it, the clock would appear to stop—because he was flying along with the light coming from the clock showing it still at noon. Does time in effect stop for someone flying at the speed of light? Einstein imagined looking at the light beam with which he was flying in tandem; it should look to him like a stationary wave of electromagnetic energy because he was not moving relative to it. But such a stationary wave would not be allowed by Maxwell's already established theory of electromagnetism. Something was wrong, Einstein concluded. He had these thoughts in 1896 when he was just 17 years old. Another 9 years would pass before he figured out how to fix the error. The resulting solution was nothing less than a revolution in physics, a revolution in our conception of time and space.

When Einstein was 4 years old, his father showed him a compass. To the boy it was a miracle—setting him on a course in science. Between the ages of 12 and 16, Einstein taught himself Euclidean geometry and differential and integral calculus. He was a bright lad, but more important, a bright lad with interesting ideas of his own. Early on, Einstein became fascinated with James Clerk Maxwell's theory of electromagnetism—the most exciting theory in science at the time. We'll look at this extraordinary theory carefully, for it is the platform on which Einstein's theory is built.

Maxwell's Theory of Electromagnetism

Scientists had long known that two types of electric charges, positive and negative, existed. For example, protons have positive charges whereas electrons have negative charges. Positive and negative charges attract each other, while negative repels negative and positive repels positive. Furthermore, scientists understood that charges can be either static or moving. Static charges have electric interactions of the sort found in static electricity. Moving charges not only have these but also magnetic interactions, as is the case when charges moving through a wire create an electromagnet.

Maxwell developed a set of four equations governing electromagnetism. In these equations, there is a constant, c, a velocity that describes the relative strengths of the electric and magnetic forces between charged particles. Maxwell devised a clever apparatus to measure c. On one side were two parallel plates, one charged negatively and one charged positively, which attracted each other due to the electric force between them. On the other side were two coils of wire with current flowing through them, which attracted each other due to the magnetic force between them. He balanced the magnetic force between the coils against the electric force between the plates to determine the ratio of magnetic and electric forces and therefore the value of c. It turned out to be approximately 300,000 kilometers per second.

Maxwell soon found a remarkable solution to his equations: an electromagnetic wave, a ripple of electric and magnetic fields, traveling through empty space at the speed c. He recognized this as the velocity of light because astronomers had already measured that.

Back in 1676, the Danish astronomer Olaus Roemer had carefully observed the satellites—moons—orbiting Jupiter. Noting that they moved around the planet like the rotating hands of an elaborate clock, Roemer saw that when Earth was closest to Jupiter, this "clock" seemed about 8 minutes fast, whereas when Earth was farthest from Jupiter (on the opposite side of its orbit), the "clock" seemed some 8 minutes slow. The difference between the two results arose because, when Earth was farthest from Jupiter, the light had to travel an additional 16 minutes to reach Earth (crossing an extra distance—the diameter of Earth's orbit—which had already been measured through astronomical surveying techniques). Roemer thus calculated that light was moving at 227,000 kilometers per second.

Then in 1728 the English astronomer James Bradley measured the speed of light by using the effect that causes vertically falling rain to appear to fall at a slant when seen from a moving vehicle. From the slightly changing deflections of starlight he observed during the year as Earth circled the Sun, Bradley deduced that the speed of light was about 10,000 times faster than the velocity of Earth as it moved around the Sun, or about 300,000 kilometers per second.

So Maxwell knew the velocity of light, and when in 1873 he calculated the speed of his electromagnetic waves and found them to be traveling at 300,000 kilometers per second, he suddenly realized that light must be electromagnetic waves. It was one of the great discoveries in the history of science. (Maxwell also understood that electromagnetic waves could have different wavelengths and predicted that some of these wavelengths could be far shorter or longer than those of visible light. Shorter ones have since been found to include gamma rays, X-rays, and ultraviolet rays, whereas longer ones correspond to infrared waves, microwaves, and radio waves. Directly inspired by Maxwell's results, Heinrich Hertz in 1888 succeeded in transmitting and receiving radio waves, which marked the invention of radio.)

Einstein's Theory of Special Relativity

Maxwell's work fascinated Einstein. But it also worried him because he had envisioned what a light beam would look like if he flew along beside it at the speed of light. According to his thinking, an electromagnetic wave would then appear stationary with respect to him—a static wave with hills and valleys just sitting like furrows in a field. But Maxwell's equations did not allow such a static phenomenon in empty space—so something had to be wrong.

Einstein noticed something else too. Suppose you move a charged particle rapidly past a stationary magnet. According to Maxwell, the moving charge would be accelerated by a *mag-*

netic force. Now take a stationary charge and move a magnet rapidly past it instead. According to Maxwell's equations, the changing magnetic field produced by the moving magnet would create an electric field, causing the charge to accelerate due to an *electric* force. The physics would be completely different in the two cases, yet the resulting acceleration of the charged particle would be identical. Einstein had a bold new idea. He thought the physics must be the same in the two cases, since the only important relationship appears to be the relative velocity of the magnet and the charged particle.

In the history of science, great breakthroughs often occur when someone realizes that two situations thought to be different are actually the same. Aristotle believed gravity operated on Earth to make objects fall toward it, but that different forces operated in the celestial realm to drive the planets and the Moon around. Newton realized that the same force that caused an apple to fall to Earth also kept the Moon in its orbit. He realized that the Moon was continually "falling" toward Earth because the straight-line trajectory it would otherwise follow in space was continually being bent to form a circle. This was not at all obvious.

Something else about light appeared very peculiar. Suppose Earth were moving through space at 100,000 kilometers per second. Wouldn't a light beam passing us in the direction of Earth's motion then go by us at only 200,000 kilometers per second (300,000 minus 100,000)? And if a light beam were traveling in the opposite direction, wouldn't that pass us at 400,000 kilometers per second (300,000 plus 100,000)? Yet light always seems to pass Earth at the same speed, regardless of direction. In 1887, physicist Albert Michelson of the Case School of Applied Science in Cleveland and chemist Edward Morley of the neighboring Western Reserve University had determined that to be true by splitting a light beam so that one half went north and one half went east. Two mirrors then reflected the beams back to their point of origin. Michelson and Morley figured that if light moves at 300,000 kilometers per second through space and their apparatus was moving through space at a speed of 30 kilometers per second (in accord with Earth's velocity around the Sun), then the speed of light relative to their apparatus would be 300,000 kilometers per second plus or minus 30 kilometers per second, depending on whether the light beam was moving opposite to or parallel with the Earth's motion. They calculated that the light beam going to and fro on a line along the direction of Earth's motion should arrive back noticeably late, compared to one going back and forth on a line perpendicular to the direction of Earth's motion. Yet their experiment showed, with high accuracy, that the two beams always arrived back at the same time.

Boy, were Michelson and Morley surprised. Knowing their apparatus was accurate, they wondered if Earth's velocity around the Sun at the time of their experiment could be canceled out by a motion of the entire solar system in the opposite direction. So they repeated the experiment 6 months later, when Earth was moving in the opposite direction in its orbit around the Sun. In that case, it should then be moving at 60 kilometers a second through space. Still the results were the same.

With all of this remarkable information at hand, in 1905 Einstein came up with two astonishing postulates. First, the effects of the laws of physics should look the same to every observer in uniform motion (motion at a constant speed in a constant direction, without turning), and second, the velocity of light through empty space should be the same as witnessed by every observer in uniform motion.

At face value, the postulates seem to contradict common sense—how can a light beam pass two observers at the same speed if those observers are moving relative to each other? Yet

Einstein proceeded to prove numerous theorems based on these two postulates, and experiments have since confirmed their accuracy many times.

Einstein proved his theorems by inventing various clever thought experiments. He called his work the theory of special relativity—*special* because it was restricted to observers in uniform motion, and *relativity* because it showed that only relative motions were important.

We should pause to admire the stunning originality of all of this. No one had ever done anything quite like it in science before. How did Einstein come to think of this? Undoubtedly, his reverence for what he called the "holy" geometry book, which he had acquired at age 12, played a role. This book described how the ancient Greek mathematician Euclid had shown that, given a few postulates defining points and lines and the relations they obeyed, one could prove numerous remarkable theorems based on them. Einstein was greatly impressed by this system: simply adopt a couple of postulates and see what you can prove. If your reasoning is sound and your postulates are true, then all your theorems should prove true as well. But why did Einstein adopt his particular two postulates?

He knew that Newton's theory of gravitation obeyed the first postulate. According to Newton's theory, the gravitational force between two objects depends on the masses of the two bodies and the distance between them, but not on how fast the bodies are moving. Newton had assumed that there was a state of rest but no way exists by gravitational experiment to find out if, for instance, the solar system is at rest or not. According to Newton's Laws, the planets would circle the Sun in exactly the same way if the solar system were stationary—at rest—or in rapid uniform motion. Einstein held that if it can't be measured, a unique state of rest simply doesn't exist. All observers moving with uniform motion could equally make the claim that they were at rest. And if gravitation can't establish a unique state of rest, Einstein thought, why shouldn't this be true for electro-magnetism as well? Based on his thinking about the charged particle and the magnet, Einstein concluded that only the rela-tive velocity of the two counted. By observing just the interac-tion between the two, someone couldn't tell whether the charge or the magnet was "at rest."

Einstein based his second postulate on the fact that Max-well's equations predicted that in empty space, electromagnetic waves would propagate at 300,000 kilometers per second. If you were "at rest," light should pass you at that speed. If you saw a light beam pass you at any other speed, that would con-stitute evidence that you were not "at rest." (In fact, Michelson and Morley had hoped to use this effect to prove the Earth was not "at rest," but they failed.) Einstein thought that all observers in uniform motion should be able to consider themselves "at rest" and should therefore always see light beams passing them at 300,000 kilometers per second. Einstein's second postulate meant that an observer traveling at high velocity and perform-ing the Michelson-Morley experiment must always fail to get a result. (Asked years later, Einstein admitted that he had known of the Michelson-Morley experiment in 1905, but claimed it did not heavily influence his thinking—he just assumed that any such effort would fail. But today, we would say that the Michel-son-Morley experiment constituted perhaps the strongest clue in 1905 that Einstein's second postulate was correct.)

Einstein figured out that light could appear to travel always at the same velocity as it passed observers traveling at different speeds relative to each other only if their clocks and rulers dif-fered. If a rapidly moving astronaut had clocks and rulers that differed from mine, then perhaps the astronaut could still measure the light beam to be passing him at 300,000 kilome-ters per second as well.

Isaac Newton had imagined a universal time that all observers could agree on, under which a moving clock would tick just as fast as a stationary one. Newton's concept of the universe is exemplified by old commando movies. Before starting the mission, the leader gathers all the members of the team together and says something along the lines of "Synchronize your watches: it's now 2:10 P.M." Everyone then sets their watches to exactly 2:10 P.M. Later, the leader counts on Newton's idea that even though the different commandos have traveled on widely different paths at different speeds (by plane, by boat, and so on), they can all get to the target at the same time. If one of them was traveling by spaceship at nearly the speed of light, however, the mission would be in trouble. A spaceship speeding past me has clocks that cannot be synchronized with mine. According to Einstein, universal time does not exist. Time is different for different observers. This opens the way for time travel.

Why a Moving Clock Ticks Slowly

One of the first theorems Einstein proved from his two postulates showed that if an astronaut were to pass me at high speed, I should see his clocks ticking slowly relative to mine. Einstein proved the idea by using a clever thought experiment: he imagined constructing a simple clock by letting a light beam bounce back and forth between two mirrors. The clock "ticks" every time the light hits a mirror. Light traveling at 300,000 kilometers per second translates into about 1 billion feet per second, or 1 foot in a nanosecond (a billionth of a second). If I separate the two mirrors by 3 feet, my light clock will tick once every 3 nanoseconds (see Figure 3). Now suppose a rocket passes me at 80 percent of the speed of light. On board is an astronaut with a light clock identical in length to mine. If I look at the astronaut's clock as it flies by, I observe the light bouncing back and forth to be traveling on a zigzag path as his pair of mirrors moves from left to right (Figure 3). As the light beam travels from his bottom mirror to his top mirror, I see the light beam traveling diago-

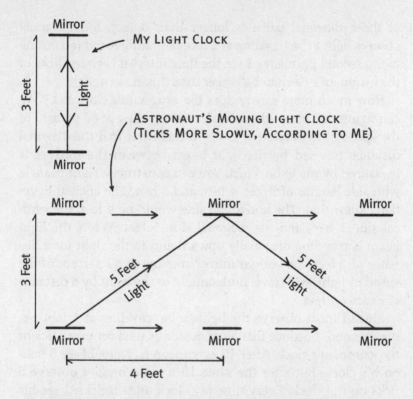

Light travels at a constant velocity of 1 foot per nanosecond.

Figure 3. Different Light Clocks

nally upward and to the right because, when the light beam arrives, I see the top mirror to the right of where it was when the light beam started. As the light beam comes back down, I see it moving diagonally downward and to the right, finally reaching the bottom mirror again, but at a position well to the right of where it was originally. The distance I measure for each of these diagonal paths is longer than 3 feet. Since I must observe light to be traveling at 1 foot per nanosecond (according to the second postulate), I see the time interval between ticks of the astronaut's clock to be longer than 3 nanoseconds!

How much more slowly does the astronaut's clock tick? We can figure this out. If the astronaut is traveling at 80 percent of the speed of light relative to me, it turns out that the diagonal distance traveled by the light beam between the mirrors is measured by me to be 5 feet. You can construct a right triangle with side lengths of 5 feet, 4 feet, and 3 feet. The ancient Egyptians knew that. The horizontal side would be 4 feet, the vertical side 3 feet, and the diagonal side 5 feet. While the light beam is traveling diagonally upward and to the right for a distance of 5 feet, the bottom mirror, moving at 80 percent of the speed of light, will travel horizontally to the right by a distance of exactly 4 feet.

Since I must observe the light to be traveling at 1 foot per nanosecond, I deduce that 5 nanoseconds pass for every tick of the astronaut's clock. After 15 nanoseconds, I should see 3 ticks on his clock. But after the same 15 nanoseconds, I observe 5 ticks on my clock. Every time my clock adds 5 ticks, I see his clock adding 3 ticks; I observe that his clock is not ticking as fast as mine.

Now for the really remarkable part. The astronaut could use his heartbeat as another kind of a clock. His parallel mirrors (with a light beam bouncing back and forth) and his heart are just two clocks at rest with respect to each other, so they should have a fixed ratio between their beats. When I look at the astronaut traveling by me at 80 percent of the speed of light, I therefore not only see his light clock ticking 3 times for every 5 times my light clock ticks, but his heart should appear to beat more slowly than mine, by the same factor. Therefore I should see him age more slowly than I do. When I aged by 5 years, I would observe that he had aged by only 3 years. Biological clocks and light clocks must slow equally; otherwise, the astronaut could tell he was moving, which would violate the first postulate.

These effects become more dramatic as the astronaut's velocity gets closer to the speed of light. The results depend on the ratio v/c, where v is the astronaut's velocity relative to me and c is the velocity of light. Can you recall high school geometry, beyond the spitballs the kid behind you used to throw? One of the theorems was Pythagoras's rule, which states that in a right triangle, the sum of the squares of the vertical and horizontal sides of the triangle equal the square of the length of the diagonal side. While I see the light moving 1 foot diagonally, the astronaut's clock is sliding to the right by a distance of 1 foot times the number (v/c), creating two sides of a right triangle. If the diagonal equals 1 foot, and the horizontal side equals (v/c) feet, then in accord with Pythagoras's rule, the vertical side equals $\sqrt{[1 - (v/c)^2]}$ feet. (Squaring the square root of $[1 - (v/c)^2]$ results in $[1 - (v/c)^2]$; adding that to $(v/c)^2$ equals 1.) The vertical progress I see the light beam making toward the upper mirror is thus not 1 foot, but 1 foot times $\sqrt{[1 - (v/c)^2]}$. Since the beam must move upward by 3 feet before I can see the astronaut's clock tick, his clock must tick at a factor of $\sqrt{[1 - (v/c)^2]}$ as fast as mine. If the astronaut passes me at a speed equal to 99.995 percent of the speed of light, I would see the astronaut's clock ticking at a rate that is one hundredth as fast as mine. After 1,000 years had passed on Earth, people on Earth would observe that the astronaut had aged only 10 years!

Time travel to the future is made possible by the fact that observers who are moving relative to each other have different ideas of time. Such observers can even disagree about which events are simultaneous—a phenomenon that will play an important role in understanding how time travel to the past may occur.

Imagine that an astronaut passes me at 80 percent of the speed of light, and I observe his rocket taking 22.5 nanoseconds to pass by me, going from left to right. At a speed of 0.8 feet per nanosecond, he moves 18 feet relative to me in those 22.5 nanoseconds, so I say his rocket is 18 feet long. I see that he is sitting in the middle of the rocket, so I say he is 9 feet from the front and 9 feet from the back. He sends light signals to mirrors at the front and back of his rocket. They reflect off the mirrors, and he receives them back at the same time. Since the mirrors are equidistant from him and the velocity of light as measured by him must be 300,000 kilometers per second, he must assert that the light signals hit the mirrors simultaneously. According to me, the light signal he sent toward the back of the rocket takes 5 nanoseconds to get there. During that time, the light signal travels 5 feet to the left and the rocket (moving at 80 percent, or $\frac{4}{5}$ths, the speed of light) travels 4 feet to the right, closing the 9-foot distance.

But what about the light signal sent by the astronaut toward the front of the rocket? It must catch up with the front, which is moving away. I see it take exactly 45 nanoseconds to catch up. The front has a 9-foot head start on the light beam. During 45 nanoseconds, the light travels 45 feet while the front of the rocket travels 80 percent of this distance, or 36 feet—so the light makes up the 9-foot starting difference. Thus, I observe that the astronaut's signal toward the back reaches the back after only 5 nanoseconds while his signal toward the front reaches the front after 45 nanoseconds. According to me, the signal aimed toward the back hits *before* the one aimed toward the front hits there.

When the signal hits the back of the rocket, it reflects off the back mirror and starts toward the front, returning to the astronaut. How long does this take, from my perspective? The light beam has to catch up with the astronaut, who has a 9-foot head start; therefore, by the argument just given, it takes 45 nanoseconds to do this. I see the signal hitting the back, bouncing off the back mirror, and reaching the astronaut again after 5 plus 45, or a total of 50, nanoseconds. What about the signal reflecting off the front mirror? It takes only an additional 5 nanoseconds to return to the astronaut because the light beam moves backward 5 feet in that time while the astronaut moves 4 feet forward to meet it. Again, the total is 50 nanoseconds for the light beam's round trip to and from the front mirror.

I see both light beams arriving back to the astronaut at the same time—just as he does. But he perceives the light beams hitting the front and back mirrors simultaneously because he is in the middle of his rocket and perceives himself as at rest. He and I disagree about whether the arrival of the light beams at the front and back mirrors are simultaneous events. It's not that one of us is right and the other one is wrong; each is right in his own frame of reference.

Now for another remarkable result. I say that the astronaut gets his light signals back after 50 nanoseconds. But I know he is moving at 80 percent of the speed of light, so I should see all his clocks ticking at 60 percent of the rate of mine. I know that his clocks should say that only 30 nanoseconds have elapsed between the time he sent the signals and the time he got them back. He thinks he is at rest, and he knows light travels at 1 foot per nanosecond, so it must appear to him that the light signals take 15 nanoseconds to reach either the front or the rear and another 15 nanoseconds to return. Thus, he concludes that the front and the back of his rocket are each 15 feet away from him and that his whole rocket must be 30 feet long. But remember that I had measured his rocket to be 18 feet long. So I say his rocket is only 60 percent as long as he measures it to be. This is the same factor that tells me how fast I see his clock ticking. Measuring sticks carried by the astronaut parallel to his direction of motion must appear to me to be compressed. If they were not, light clocks carried by the astronaut parallel and perpendicular to his direction of motion would tick at different rates—and he could tell he was moving, which is not allowed by the first postulate.

The preceding discussion has been aimed at the left hemisphere of your brain (the verbal-logical one). I'm now going to explain it to your right hemisphere (the visual-spatial one). Figure 4 shows a spacetime diagram of the previous discussion. The astronaut's world line and the world lines of the front and back of his rocket are shown as lines tilted to the right. Take a ruler, hold it horizontally, and scan it from the bottom to the top of the diagram. The intersection of the world lines with the ruler will show a movie of how the scene plays out from my point of view. Watch the astronaut and his rocket move from left to right with time as you move the ruler slowly upward. The light beams he sends out to the front and back of the rocket and receives back are shown as 45-degree lines, since I must observe them traveling at 1 foot per nanosecond. The arrival of the light beam at the back of the rocket (event A) I see occurring before the arrival of the light beam at the front of the rocket (event B). But the astronaut, who thinks he is at rest, sees the two events as simultaneous (occurring at 15 nanoseconds astronaut time, as indicated by the little clocks and the slanted line labeled *15 ns AT* connecting them).

A similar experiment performed by an earthling is also shown in Figure 4. The earthling's world line goes straight up because, according to me, he is not moving with time as I scan upward with the horizontal ruler. The line connecting the earthling's clocks showing 15 nanoseconds earthling time (labeled *15ns ET*) is horizontal because he is at rest with respect to me.

Spacetime is like a loaf of bread, set on end. If I slice the bread horizontally, I will get slices that represent different in-

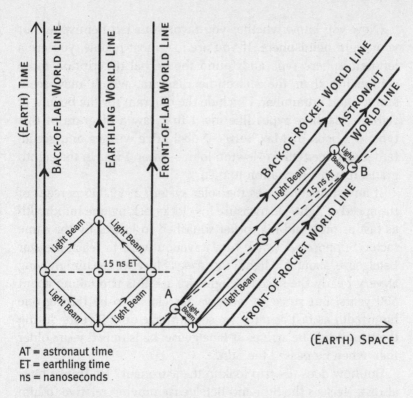

AT = astronaut time
ET = earthling time
ns = nanoseconds

Figure 4. Different Perceptions of Time: The Astronaut and the Earthling

stants of Earth time. Two events are simultaneous if they are in the same slice. But a moving astronaut will slice the loaf differently, on a slant, the way French bread is cut. Events that are in the same slanted slice will appear simultaneous to the astronaut. This also explains why the astronaut and I can disagree on the width of his rocket. He and I are just slicing its four-dimensional world line differently. It's like asking how wide a tree trunk is. Saw it horizontally, and you will get one answer; saw it on a slant and you will get a different one.

(Now you know whether you favor your left hemisphere or your right hemisphere. If you are like most people, you are a left-hemisphere type and found the verbal description more compelling than the spacetime diagram, which may seem strange and unfamiliar. I include the diagram for the benefit of right-hemisphere types like me. I first saw a diagram of this type in a book by Max Born—Nobel Prize winner, and grandfather of singer Olivia Newton-John—when I was in the eighth grade. It was a revelation to me.)

If an astronaut passes the solar system at 99.995 percent of the speed of light, we measure his clocks ticking one hundredth as fast as ours and his rocket squashed in length by the same factor. Suppose we watch him travel outward to reach the star Betelgeuse, about 500 light-years away. Since we see him moving at very nearly the speed of light, we see this trip taking about 500 years. But since we observe his clocks to be ticking one hundredth as fast as ours, we see him age only 5 years during the trip. When he arrives at Betelgeuse, he is only 5 years older than when he passed the Sun.

But how does the trip look to the astronaut? He thinks he is at rest. He sees the Sun and Betelgeuse moving relative to him at 99.995 percent of the speed of light, so he measures them to be only 5 light-years apart (one hundredth of the 500-light-year separation that we measure). The Sun and Betelgeuse are like the front and back ends of a "rocket" passing him at nearly the speed of light. He measures its total length to be 5 light-years. Thus, traveling at nearly the speed of light, the back end of the "rocket," Betelgeuse, passes him about 5 years after the Sun does, so when he encounters Betelgeuse, he is only 5 years older, just as we would expect.

Interestingly, Einstein's thought experiments did not involve people on Earth looking at an astronaut going by on a rocket; rather, Einstein analyzed the case of an observer at a train station comparing notes with an observer riding in the middle of a rapidly moving train. Einstein used a train because that was the fastest vehicle that had been created by 1905.

Why Rockets Can't Go Faster Than Light

If an astronaut's rocket were to travel by us at faster than the speed of light, a light beam he sent forward could never catch up with the front of his rocket. The light beam could never catch up because the front of the rocket would be moving faster and have a head start. Any athlete should know that catching another runner who is running faster and has a head start is impossible. The astronaut's observations would be most peculiar: he would take out a flashlight and shine it toward the front of his rocket, but he would never see that beam of light arrive. That's not what an observer at rest would see: rather than perceiving he was at rest, this astronaut would *know* he was moving—and that's not allowed by the first postulate.

Thus, Einstein concluded, nothing can travel faster than the speed of light. He had discovered a speed limit for the universe—the velocity of light. It's written into the fabric of the universe, right in the equations of electrodynamics. This speed limit results directly from Einstein's two postulates, which we trust because many results derived from them have been checked. In our largest particle accelerators, we speed up protons. We keep pushing them harder and we observe them going faster and faster, approaching the speed of light but never getting to it—just as Einstein predicted.

$E = mc^2$ is another result Einstein proved from the two postulates. (Here E is energy, m is mass, and c^2 is the speed of light squared.) Of course, the speed of light is a very large quantity—and it is squared in the formula—so the loss of just a little mass is accompanied by release of an enormous amount of energy. When an atomic bomb goes off, a small amount of mass is converted into an enormous amount of energy. The atomic bomb works. The postulates seem to work. And so we don't expect to see an astronaut flying by us at a speed faster than the speed of light.

A Universe of Four Dimensions — or More

Our universe is four-dimensional—there are three dimensions of space and one dimension of time. H. G. Wells thought the dimension of time was just like one of the dimensions of space, but he was wrong. There is a crucial difference between them. It turns out that mathematically a minus sign is associated with the dimension of time. This one little minus sign makes all the difference: separating the future from the past, allowing causality in our world, and making it difficult to travel freely in time. To explore the idea of time travel, we therefore have to understand how that minus sign arises. And that, in turn, requires considering what moving observers can agree on, since there's so much they can disagree about.

First, when comparing separations in space and separations in time, we must use units for which the speed of light is 1. Light-years and years are such units. Light travels at a speed of 1 light-year per year. We could equally well use units of feet and nanoseconds, for light travels at a speed of 1 foot per nanosecond. The speed of light is a "magic" velocity, a velocity everyone can agree on, so it can be used to compare separations in space with separations in time.

Consider the previous example. An astronaut passes me at 80 percent of the speed of light. He sends light signals to the back and front of his rocket where they reflect off mirrors and return to him. I see his sending and receiving of the signals to be two events separated by 40 feet in space and 50 nanoseconds in time. Meanwhile, the astronaut, who perceives himself to be at rest, sees the two events separated by zero feet in space and 30 nanoseconds in time as measured by his clocks. We disagree about the separation of the events in both space and time.

But we can agree on the square of the separation in space *minus* the square of the separation in time. If I take the square of the separation in space I measure (in feet) and subtract the square of the separation in time I measure (in nanoseconds), I get $40^2 - 50^2 = 1,600 - 2,500 = -900$. If he takes the square of the separation he measures in space (in feet) and subtracts the separation he measures in time (in nanoseconds), he gets $0^2 - 30^2 = -900$.

We both get the same answer. If this quantity is negative, we say that the two events have a *timelike separation:* these events have a greater separation in time than they do in space, making the result negative. When this quantity is positive, we say that two events have a *spacelike separation.* Such events have a separation in space larger than their separation in time. And when this quantity is zero, we say the two events have a *lightlike separation:* they are two events that may be connected by a light beam. The astronaut and I would agree that two such events have an equal separation in space and in time. We may differ on how many feet and nanoseconds separate them (I may say 5, and he may say 15), but we both agree that the two numbers are equal. That's because by Einstein's second postulate we must both observe the light beam connecting the two events to be traveling at a speed of 1 foot per nanosecond. All observers agree on the quantity of "the square of the separation in space minus the square of the separation in time," ensuring that they will all agree that the speed of light is 1 in these units, regardless of how their measuring sticks and clocks differ. The minus sign assures us that all observers will agree on the speed of light.

Suppose you are invited to a party 5 years from now on Alpha Centauri, which is 4 light-years away. You can get there by rocket, traveling at 80 percent of the speed of light. All observers would agree that the party is to the "future" of where you are now because you can plan now to attend it. That party is separated from "you-now" by a distance of 4 light-years in space and 5 years in time. So, using light-years and years as our units, we measure the square of the separation in space minus the square of the separation in time to be $4^2 - 5^2 = 16 - 25 = -9$. The party has a timelike separation from you-now. Any two such events can be connected by a rocket ship that travels between them.

But a gala 3 years from now on Alpha Centauri is an event you cannot get to because you can't travel faster than light. That event has a spacelike separation from you-now (it's in your present). The square of its separation in space minus the square of its separation in time is positive: $4^2 - 3^2 = 16 - 9 = 7$. An observer traveling at 75 percent of the speed of light toward Alpha Centauri would assert that you-sitting-on-Earth-now and that gala on Alpha Centauri are simultaneous events. He would not be surprised that you could not attend. How could you? According to him, both events occur at the same time.

Now consider a celebration on Alpha Centauri 6 years ago. That event is in your "past." An astronaut could have attended that celebration and be here now. He could have traveled here at two thirds the speed of light. (The celebration and you-now have a timelike separation so the astronaut can visit you-now after having attended the celebration. The celebration is therefore in the "past" of where you are now.) Thus you may divide our four-dimensional universe into three regions—the past, the present, and the future.

We can picture this in a 3-D diagram showing two dimensions of space horizontally and the dimension of time vertically, with the direction of the future as up and the direction of the past as down (see Figure 5). We perceive the Sun as at rest, so from our viewpoint we will show the Sun's path through spacetime as a vertical world line. The star Alpha Centauri is another vertical world line, 4 light-years away. The diagram shows light waves emitted by you—say, for purposes of illustration, in the year 2001—traveling to the future, tracing out what is called the *future light cone*. As Stephen Hawking has noted, these light waves spread out like circular ripples in a pool at a speed of 1 light-year per year. If we want to see what the universe looks like at a given time, we just cut a horizontal slice out of our 3-D diagram and look at it. A horizontal slice will cut the future light cone in a circle. At a given instant, the emitted light waves look like a circle around you. Cut a horizontal slice later, and the circle is larger. Pan upward, and the light cone's diameter grows larger and larger. Since the light moves horizontally outward by 1 unit (1 light-year) for every unit upward in time (1 year) it goes, the cone has an angle of 45 degrees in the diagram. The world line of Alpha Centauri pierces this light cone (in 2005). You can send a signal to any event lying on the future light cone. The party on Alpha Centauri 5 years from 2001, in 2006, lies inside the future light cone. That event is in your future. You are able to visit events inside the future light cone.

Also shown is the *past light cone*, a contracting cone arriving at you-now. Events on the past light cone are events you can see today. The past light cone intersects the world line of Alpha Centauri 4 years ago—in this case, in 1997. Light beams from Alpha Centauri emitted in 1997 arrive here in the year 2001. When you look at Alpha Centauri now, you see it as it appeared 4 years ago. The farther away you look, the farther back in time you see. The view of the universe you have is the past light cone. Everything inside this past light cone is now in your "past"; it encompasses events that you could have visited (for instance, in this case, the celebration on Alpha Centauri in 1995 —attending it would still have left you enough time to reach your current location on Earth in 2001). Since the speed of light

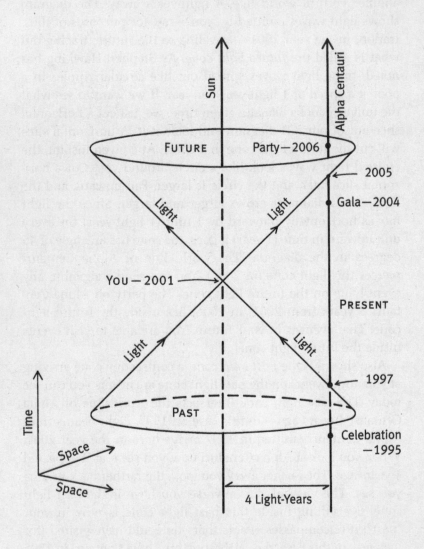

Figure 5. Past and Future Light Cones

through empty space is the upper speed limit in the universe, no events *outside* the past light cone could have had any influence on you yet. In between the future light cone and the past light cone lies the "present." It includes events that someone on some rocket ship may think are simultaneous with you-now. That gala on Alpha Centauri in 2004 is in your "present." Even though you, sitting still on Earth in the year 2001, think that it has yet to occur, other observers think that it is occurring simultaneously with you-now and still others (on even faster rockets) think it has occurred before you-now. This gala is thus in your "present." Notice that the "future" and the "past" —like the top and bottom of an hourglass—are two separate pieces touching only at the single event "you-now." The "present" circles around these two cones and consists of one connected piece. All observers agree on which events are in each region (the past, present, and future of the event you-now) because all observers see light traveling at the same speed, and all observers agree on which side of each light cone a given event resides.

Since you can't travel faster than the speed of light, your future world line must lie entirely within the future light cone. Your world line can never make an angle larger than 45 degrees with respect to the vertical, for that would mean you were traveling faster than light. Thus, your world line must, like the helical world line of Earth in Figure 1, always proceed toward the future. This prevents you from circling back to the past as the time traveler does in Figure 2. Any world line that completes a circle in spacetime like that must, at some point, be tipped at an angle larger than 45 degrees with respect to the vertical. Time travel to the past would therefore imply exceeding the speed of light at some point, which is not allowed by special relativity. (Later, when we consider time travel to the past, we will discuss a way around this seemingly insurmountable difficulty.)

Flatland and Lineland

That we have three dimensions of space and one dimension of time is interesting. We might, for example, have ended up with a universe with just two dimensions of space and one dimension of time. This would be the world of Flatland, as described by Edwin Abbott in a wonderful book from 1880 and updated by A. Dewdney in his book *Planiverse*. Creatures living in Flatland would be able to move but only in two spatial dimensions, "up-down" and "left-right." Flatlanders would find life somewhat different from ours. A Flatlander could have a mouth and

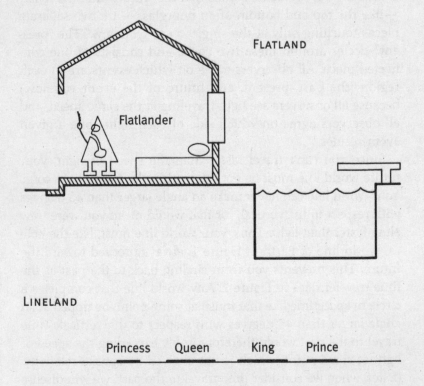

Figure 6. Flatland and Lineland

a stomach—but no alimentary canal passing all the way through his body, for then his body would fall apart into two pieces! (See Figure 6.) Flatlanders would have to digest their food, then throw up after dinner, a peculiarity noted by Hawking in *A Brief History of Time*. A Flatlander could see with a circular eye with a retina on the back. He could read a newspaper that would be a line, with Morse code dots and dashes printed on it. He could have a house with a door and a window. He could even have a swimming pool in the back yard. But he would have to climb over the roof of his house to get to his back yard. And he would have to fall backward into bed. Life in a universe with two spatial dimensions and one time dimension would be more constricted than our own.

A world with only one dimension of space and one dimension of time—Lineland—would be even simpler. Creatures in Lineland would be line segments (see bottom of Figure 6; to make them visible, these Linelanders are drawn thicker than the line; actually both they and the line are of zero thickness). There could be a king and queen of Lineland. The king might be to the right of the queen. There could be a prince to the right of the king and a princess to the left of the queen. If the queen was on your left, she would always remain on your left—she could never move around you to get to your right side. In Lineland, left and right would have an absolute separation, just like the separation between the future and the past.

How Many Dimensions?

The reason we have three dimensions of space and one dimension of time may lie in how gravity works. Einstein explained gravity by showing how mass causes spacetime to curve. When one generalizes Einstein's theory of gravity into spacetimes with various dimensions, one finds that massive objects in Flatland do not attract each other—there is no gravitational attraction at a distance (nothing to keep our Flatlander's water in his pool). Therefore, large objects might not assemble themselves, and thus intelligent life might not develop. (Of course, intelligent life in Lineland also seems impossible.) But with three dimensions of space and one dimension of time, planets have stable orbits around their suns. If we had more than three dimensions of space and still one dimension of time, the orbits of planets around their stars would be unstable—again creating unfavorable conditions for intelligent life.

Suppose we had two dimensions of time. After all, ancient Aboriginal wisdom told of a second time—the dreamtime. In that case, the universe would be five-dimensional. The quantity agreed upon by moving observers would then have to be the sum of the squares of the separations in the three space dimensions *minus* the square of the separation in (ordinary) time *minus* the square of the separation in dreamtime. (There would be a minus sign in front of the dreamtime term also.) Because the sign of the terms connected with time and dreamtime are the same, we could rotate in the time-dreamtime plane just as we can rotate in the horizontal plane formed by the left-right and front-back dimensions. That would make time travel to the past easy. You could visit an event in your own past simply by traveling—that is, having your world line circle around—in the dreamtime direction (without ever exceeding the speed of light). If time is one-dimensional, then, like an ant trapped on a line, you can only go forward. But if there were two time dimensions (time and dreamtime), you could circle around in the time-dreamtime plane and visit anywhere in time you wanted, like an ant free to roam on a sheet of paper. Normal causality would not exist in such a world. Apparently, we do not live in such a world.

But wait. Our universe may have more dimensions than we might at first think. In 1919 Theodor Kaluza discovered that if one generalized Einstein's theory of gravity into a world with four dimensions of space and one dimension of time, one would obtain normal Einstein gravity *plus* Maxwell's equations for electrodynamics as updated by Einstein's theory of special relativity. Electromagnetism was just due to the action of gravity in an extra spatial dimension. But since we don't see an extra dimension of space, the notion sounded crazy. Then in 1926 Oskar Klein (not to be confused with Felix Klein who invented the Klein bottle, a three-dimensional version of the Möbius strip) worked out the idea that the extra dimension could be curled up like the circumference of a soda straw.

A soda straw has a two-dimensional surface; you can make a straw by cutting a vertical strip of paper and taping the left and right sides together to form a narrow cylinder. To locate a point on the straw, one needs two coordinates: the vertical position along the length of the straw and the angular position around its circumference. Creatures living on the surface of a soda straw would really be inhabitants of a two-dimensional Flatland universe, but if the circumference of the straw was small enough, their universe would look to them like Lineland. Klein proposed that a fourth spatial dimension could be curled up just like the circular dimension of a soda straw, with a circumference so small (8×10^{-31} cm) that we would not notice it.

In this universe, positively charged particles like the proton would circulate counterclockwise around the straw, whereas negatively charged particles like the electron would circulate clockwise. Neutral particles (like the neutron) would not circle the straw. The wave nature of particles would allow only an integer number (1, 2, 3, 4, and so on) of wavelengths to wrap around the tiny circumference, and therefore charges should come in multiples of a fundamental charge like those carried by the proton and electron. Kaluza-Klein theory unified the forces of gravity and electromagnetism, explaining both in terms of curved spacetime—a step toward Einstein's goal of a grand unified field theory to explain all the forces in the universe. But Kaluza-Klein theory provided no new predictions of effects that could be checked by experiment and so remained in a physics backwater.

Recently, however, superstring theory has revived the idea of extra dimensions. It proposes that fundamental particles, such as electrons or quarks, are actually tiny loops of string with circumferences of order 10^{-33} centimeter. Superstring theory (or M theory, as it is sometimes called) suggests that our universe actually has 11 dimensions—with one macroscopic dimension of time, three macroscopic dimensions of space, and seven curled-up dimensions of space each of order 10^{-33} centimeters in circumference. One of the extra dimensions could explain electrodynamics, à la Kaluza-Klein theory, while the others could explain the weak and strong nuclear forces, which are responsible for some types of radioactive decay and for binding the atomic nucleus together. Just as every position along the vertical dimension of a soda straw is not a point but rather a tiny circle, every position in space in our universe would represent not a point but a tiny, complicated, seven-dimensional space 10^{-33} centimeters in circumference. The exact shape of this space, whether like a higher dimensional sphere or donut or pretzel, would determine the nature of the particle physics we observe.

In the very early universe, our familiar three dimensions of space could have been microscopic as well. Since then, they would have greatly expanded in size and would continue to do so today, explaining the observed expansion of the universe. Why did just three dimensions of space expand, leaving the other seven curled up and tiny? As Brian Greene explains in his 1999 book *The Elegant Universe*, Brown University physicist Robert Brandenberger and Harvard physicist Cumrun Vafa have proposed that the curled-up dimensions remain tiny because they are wrapped with string loops like rubber bands around a soda straw. Brandenberger and Vafa have proposed scenarios under which collisions between string loops might usually cause three spatial dimensions to become unwrapped, allowing them to expand to large size. If fewer or more than three dimensions of space were to expand, this could create Lineland, Flatland, or even four- to ten-dimensional-land macroscopic universes, all exhibiting different laws of microscopic physics.

In such an ensemble of universes, we would expect to find ourselves in one where intelligent life could flourish—just as we find ourselves living on a habitable planet when many planets are not. This argument, called the *strong anthropic principle* by British physicist Brandon Carter, is a self-consistency argument. Given that you are an intelligent observer, the laws of physics in your universe, at least, must allow intelligent observers to develop. As intelligent observers, we might naturally find ourselves in a universe with three macroscopic dimensions of space, even though Lineland or Flatland or Higher-Dimensional-Land universes could still be out there somewhere as well.

There has even been some discussion that one of the extra curled-up dimensions proposed by superstring theory might be timelike—such as dreamtime. What would a circular extra dimension of time be like? If you headed off in the dreamtime dimension, you would keep returning to the time you started, as Bill Murray's character did in the movie *Groundhog Day*, in which he kept experiencing the same day over and over. The time-dreamtime plane would resemble a soda straw with ordinary time going up its length and dreamtime circling it. The dreamtime circumference would be about 5×10^{-44} seconds. Just as an ant crawling along a straw might make a U-turn by turning in the narrow dimension of the straw, an elementary particle might make a U-turn in ordinary time to go backward toward the past by taking advantage of the dreamtime direction to make the turn. Indeed, as I discuss later, one may interpret a positron as an electron going backward in time. This is presumably the mechanism used in the 2000 movie *Frequency* for a son to send signals—in this case, radio wave photons—to the past to save his father (physicist Brian Greene even does a cameo in the film to hint at the physics involved). I would emphasize, however, that the idea that one of the extra curled-up dimensions could be timelike (like the dreamtime) is not the standard picture.

In its standard formulation, superstring theory suggests that different universes may exist with different numbers of macroscopic spatial dimensions (up to ten) but affirms that there is always just one dimension of time. It's the only dimension marked as different—the one with the minus sign. Time thus appears unique in the laws of physics and, as Einstein showed, uniquely paradoxical.

THE TWIN PARADOX

The shortest distance between two points in space is a straight line. If you go out of your way to visit a friend en route to a party, more mileage will be recorded on your car's odometer than if you had gone to the party directly. But because of that minus sign associated with the time dimension, the situation is different when traveling between two events separated in time. If you are invited to a party 10 years from now on Earth, the straightest path to it—namely, just staying home on Earth and waiting—is the one that will cause the most ticks on your clock, 10 years' worth. If, however, you decide to make a detour and flit out to Alpha Centauri and back on your way to that party, you will move your light clock back and forth (out to

Alpha Centauri and back), stretching the distance its light beams have to travel, thus causing it to tick less. Because space and time have opposite signs, extra distance traveled in space on the way to that party means less time elapsed on your clock. You age less. This leads to the famous "twin paradox," which is the key to time travel to the future.

Suppose there are twin sisters—Eartha and Astra. Eartha stays on Earth. Astra travels by rocket at 80 percent of the speed of light to Alpha Centauri. Since Alpha Centauri is 4 light-years away, Astra's trip there will take 5 Earth years. Eartha will see Astra's clock ticking slowly—at 60 percent the rate hers ticks, so Astra will age only 3 years during the trip. Astra turns around when she reaches Alpha Centauri and returns to Earth at 80 percent of the speed of light as measured by observers on Earth. The return trip also takes 5 Earth years, so Eartha is 10 years older when Astra arrives back home. On the return trip, Eartha sees Astra's clock again ticking at 60 percent the rate of hers, so Eartha sees Astra age another 3 years during her return. When Astra gets back, she will be 6 years older, but Eartha will be 10 years older. Astra has time-traveled 4 years into the future.

Here is the paradox. Astra might argue that, according to her observations, it was Eartha who moved at 80 percent of the speed of light, instead of herself, and therefore expect Eartha to be the younger one when the two met again.

Here is what is wrong with that argument. The two sisters have not had equivalent experiences. Eartha, who stays on Earth, is an observer who moves at constant speed without changing direction. (I am ignoring the tiny velocity of Earth around the Sun.) Eartha is, therefore, an observer who satisfies Einstein's first postulate. But Astra is not an observer moving at constant speed without changing direction. She changes direction. In order to turn around when she reaches Alpha Centauri, she must slow down from 80 percent of the speed of light to zero speed and then accelerate back up to 80 percent of the speed of light in the opposite direction. Astra's world line bends, whereas Eartha's is straight. Astra, an observer who experiences acceleration and braking, does not satisfy Einstein's first postulate. When Astra brakes to a stop and reverses direction at Alpha Centauri, all her unbolted belongings fly up against the rocket wall facing away from Earth. Her stuff gets broken. (In fact, the acceleration is so powerful that, in practice, she could be killed—but for purposes of this argument we will assume she is constructed sufficiently sturdily to withstand it.) Astra would know she had turned.

When Astra is traveling outward at 80 percent of the speed of light, before she has turned, she can think of herself as being at rest. It's true that she would see Eartha's clock ticking more slowly than hers; when she is 3 years older upon arrival at Alpha Centauri, she does think that Eartha has aged only 1.8 years back on Earth. Astra reckons that her arrival on Alpha Centauri and Eartha's being 1.8 years older on Earth are simultaneous events connected by a diagonal "French-bread" slice through spacetime. This slice is tilted because Astra is moving (just as in Figure 4, where the line *15 ns AT*, astronaut time, was tilted because the astronaut was moving). Remember, Eartha and Astra would disagree about whether light beams Astra emits arrive simultaneously at the front and back of her rocket. They must likewise disagree on whether more widely separated events are simultaneous. So just before Astra arrives at Alpha Centauri, she and Eartha both will think that the other has aged less.

But now Astra reverses her direction of motion and begins slicing spacetime on a different slant. Moving at 80 percent of the speed of light *toward* Earth, she thinks that her departure from Alpha Centauri is simultaneous with Eartha-on-Earth 8.2

years from the start. On the return trip, at constant speed then, Astra would perceive that Eartha ages just 1.8 more years, from 8.2 years to 10 years. During this period Astra ages another 3 years, making her a total of 6 years older when she arrives back. Astra observes Eartha to be 10 years older to her 6 years older. There is no paradox. Astra's idea of what events are occurring simultaneously on Earth just happens to change radically when she turns around at Alpha Centauri. Astra accelerates—Eartha doesn't. Astra turns—Eartha doesn't.

The twin who goes out of her way—who has to accelerate— is the one with fewer ticks on her clock. In this case, the straight trajectory, the one Eartha takes, is the lazy trajectory. The twin who exerts herself ages less. It's almost like saying "exercise is good for you." Astra's light clock is moving back and forth, stretching the distance the light beams had to travel, causing it to tick less.

Special relativity has many results that originally appear paradoxical, yet careful consideration shows that all such paradoxes can be resolved. In this case, when the sisters meet again, they both agree that Astra aged less. Einstein's universe is not the common sense one we might first think of—but it is the universe we live in.

The twin paradox enables you to travel to the future.

A Time Machine for Stay-at-Homes

In H. G. Wells's *Time Machine*, the Time Traveler didn't board a rocket and blast off to the stars; he traveled to the future merely by sitting in his time machine at home. Such a time machine is also possible. First, disassemble the planet Jupiter and use its material to construct around yourself an incredibly dense spherical shell whose diameter is just a bit larger than the critical diameter needed for that mass to collapse to a black hole (for a

Jupiter-mass shell, that is a bit bigger than 5.64 meters, roomy enough for you to sit inside). Interestingly, Newton showed that a spherical shell of matter would exert no gravitational effects inside, a result that happens to be true in Einstein's theory of gravity as well. The bits of mass in the shell would completely surround you, and the forces they exert on you would act in all different directions, canceling each other out exactly, leaving a zero net effect. So even though the spherical shell is quite massive, once inside no gravitational forces would affect you. If you sat just outside the spherical shell you would be torn apart by tidal forces generated by its gravitational attraction, but inside the sphere you would be safe. In Einstein's theory of gravity, tidal forces are produced by a curvature, or warping, of spacetime. Outside the time machine, spacetime would be dramatically curved, but inside, where there are no forces, it would be flat (see Figure 7; in this diagram we are illustrating just two [curved] spatial dimensions instead of three, so the spherical shell surrounding you is shown as a circle). To get into your time machine without being killed, you must construct a very large spherical shell about the size of Jupiter slowly around you, to minimize the tidal forces on you during this construction process. After construction, you then adjust the forces being exerted on the shell so it slowly compresses around you.

How can this machine take you to the future? Einstein showed in 1905 that photons (particles of light) have energies that are inversely proportional to their wavelength: short-wavelength photons (as in X-rays) pack a large punch of energy whereas long-wavelength photons (as in radio waves) carry just a little. Inside your shell you are like a child trapped at the bottom of a well (see Figure 7 again). Imagine placing a heavy metal ring on a flexible rubber sheet. It would drag the rubber sheet downward until it looked like Figure 7. Ants could play on the flat rubber surface inside the ring, but if they wanted to

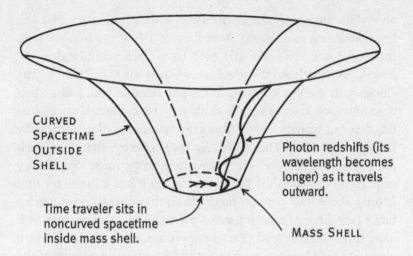

CURVED
SPACETIME
OUTSIDE
SHELL

Photon redshifts (its
wavelength becomes
longer) as it travels
outward.

Time traveler sits in
noncurved spacetime
inside mass shell.

MASS SHELL

Time traveler ages less than observers sitting outside.

Figure 7. Time Machine for Visiting the Future

leave, they would have to expend energy climbing up the
curved rubber surface outside. Likewise, you are safe at the
bottom of a "gravitational" well but climbing back out to a large
distance away from the shell would require a lot of energy,
because you would be fighting against the shell's gravitational
attraction the whole way.

If you emit a photon inside your shell, and it passes through
a window in the shell, it will lose energy as it climbs out of the
gravitational well. Distant observers will see that the photon
has less energy and, therefore, according to Einstein's 1905
paper, a longer wavelength when they detect it. The photon has
become *redshifted*—pushed toward the red, longer-wavelength
end of the spectrum. Suppose you make a clock with an electric
circuit that oscillates 1 billion times per second. This produces

an oscillating electromagnetic wave with a frequency of 1 billion vibrations per second. Traveling at 1 foot per nanosecond, this wave has a wavelength of 1 foot. Each additional wavelength emitted from the clock represents another "tick" of the clock, with the clock ticking once per nanosecond. But as this electromagnetic wave travels outward, it must climb up against the force of gravity out of the gravitational well produced by the spherical shell. This climbing takes energy; therefore, each photon, or packet of electromagnetic energy, must lose energy as it travels outward. If the spherical shell has a diameter that is only about 6.67 percent larger than that necessary to form a black hole (in this case, 6 meters across), then each photon will lose three quarters of its energy on the way out. Distant observers will find that each photon has only one quarter of the energy it had when it was emitted. A photon with one quarter as much energy has a wavelength 4 times longer. That means that distant observers would see photons with a wavelength of 4 feet passing them. Since these photons pass them at the speed of light, 1 foot per nanosecond, it would take 4 nanoseconds for 1 wavelength to pass by. The distant observers would detect electromagnetic waves oscillating once every 4 nanoseconds. Therefore, they would see the time traveler's clock ticking once every 4 nanoseconds, or 4 times more slowly than the traveler experienced. Consequently, they would see him aging 4 times more slowly than normal. After observing the time traveler for 200 years, they would see him age only 50 years.

What would the time traveler observe? Photons emitted by distant observers would fall onto his shell. These photons would pick up energy as they fell, like any falling object. When these photons hit the window of his shell and entered it, they would have 4 times as much energy as they had when they were emitted. If these photons had a wavelength of 1 foot when they were emitted, the time traveler would observe them to have a wavelength of one fourth of a foot. He would see them oscillate once every fourth of a nanosecond as they went by him rather than the once-a-nanosecond oscillation they had when they were emitted. The time traveler would thus observe the distant observers' clocks to be ticking 4 times too fast. The time traveler would see the history of the outside universe passing in front of his eyes at 4 times the normal rate. It would resemble a movie playing at fast forward. A new day's broadcast of the evening news would appear every 6 hours to the time traveler.

The traveler and the distant observers would both agree that he was aging 4 times more slowly than observers far outside. As pointed out by astronomer Thomas Gold of Cornell, the time traveler and distant observers outside age differently because their situations are not symmetrical: the time traveler is deep within a gravitational well, while they are not.

The time traveler's perspective would be like that described by H. G. Wells. He would see a candle outside his time machine burn out quickly, but that candle would look blue-white hot to him, rather than red, because photons falling into the time machine are blueshifted toward the blue, shorter-wavelength end of the spectrum. In fact, many of the photons the candle emitted would be shifted into the ultraviolet region.

After aging 50 years, the time traveler could expand the spherical shell surrounding him to a large size and then disassemble it. He would step out of his time machine only 50 years older, but 200 years would have passed outside.

(Note that a 2-solar-mass time machine of this type, 12.6 km across, would be easier to compress, so more practical to build.)

If you want to travel even faster into the future, just contract your sphere slightly, bringing it even closer to the critical size necessary to form a black hole. But there is a limit. The trouble, as explained by physicists Alan Lightman, Bill Press, Richard Price, and Saul Teukolsky in their 1975 problem book on relativity, is that even with the strongest possible material, there is a limit on how small you can make a self-supporting shell without its collapsing—the shell must have a diameter at least 4 percent larger than that required to form a black hole. In this case, the time traveler would age 5 times more slowly than those outside. Thus, there is an upper speed limit of 5 years-per-year for how fast the time traveler can go toward the future in this particular type of time machine. You don't want to get too close to this speed limit; if your shell collapses, it will form a black hole. The shell will then inexorably collapse to a size smaller than an atomic nucleus, crushing you inside. Therefore, this type of time machine is fine if you don't want to leave the solar system, if you are especially curious about what it will be like a couple of centuries into the future, and if you are willing to wait 50 years to find out.

Electrostatic repulsion of like charges could hold up a shell of matter closer to the critical radius, allowing faster travel to the future, but the machine's mass would have to be enormous, more than 20 million solar masses. (Otherwise, the enormous electric fields produced outside the sphere would create electron-positron pairs that would rapidly bleed off the charge and precipitate a collapse.) You couldn't put such a massive shell in the solar system without wreaking havoc. Likewise, time would pass slowly for you if you simply hovered just outside a black hole, but the black hole would also have to be enormous for you to survive—not suitable for inside the solar system.

Clearly, you can visit the future by staying at home, but it's still easier to do it by traveling in space.

TODAY'S TIME TRAVELERS

The *Tao Te Ching*, attributed to Lao-Tzu, says, "a journey of a thousand miles must begin with a single step." The Wright brothers' first flight went only 120 feet. The first radio transmission crossed just a single room. We should realize that we do have time travelers with us even today. They have already made that first step.

Astronauts experience the effect of aging a little less than the rest of us. Because the Russian cosmonaut Sergei Avdeyev was in orbit a total of 748 days during three spaceflights, he's about one fiftieth of a second younger than he would be if he hadn't gone on those trips. This results from the interaction of two effects. First, a clock sitting at rest with respect to Earth at the altitude of the Mir space station would tick slightly faster than one on Earth's surface. That's because Mir is higher up in Earth's gravitational well. But the second and larger effect is that the astronaut would be traveling at more than 17,000 miles per hour, and his clock would be ticking more slowly than if he were stationary with respect to Earth's surface. His orbital velocity is 0.00254 percent of the speed of light—the slowing of his clock is small, but real.

Mr. Avdeyev is our greatest time traveler to date. Other astronauts have also traveled to the future. For example, Story Musgrave, who helped repair the Hubble Space Telescope, has spent a total of 53.4 days in orbit: he is thus more than a millisecond younger than he would be if he had stayed home. Astronauts going to the Moon traveled faster than Mr. Avdeyev, but their trips lasted just a few days, making the total distance they were displaced in time smaller. Mr. Avdeyev has traveled to the future by about one fiftieth of a second. That's not much, but it is a step. A journey of a thousand years must begin with a fraction of a second.

3 TIME TRAVEL TO THE PAST

> There was a young lady called Bright
> Who could travel far faster than light;
> She set off one day,
> In a relative way,
> And returned home the previous night.
> — A.H.R. BULLER

YOU CAN SEE THE PAST

If you want only to see the past, rather than visit it, then you have an easy task. We are already doing it today—because of the finite velocity of light. If we observe Alpha Centauri, 4 light-years away, we see it not as it looks today but as it looked 4 years ago. The star Sirius, 9 light-years away, we see as it shone

9 years ago. If you look at the Andromeda galaxy, 2 million light-years away, you see it as it appeared 2 million years ago—at a time when our "grandparent" species, *Homo habilis*, walked the Earth. We see the more distant Coma cluster of galaxies as it appeared 350 million years ago, when amphibians had just crawled out of Earth's seas. The quasar 3C273 is over 2 billion light-years away, so we see it at an epoch when the most complex life forms on Earth were bacteria. (Quasars are bright objects probably powered by gas falling toward massive black holes in the centers of galaxies.) A distant quasar, discovered recently by my Princeton colleagues Michael Strauss and Xiaohui Fan, is over 12 billion light-years away.

As we look farther away, we look farther back in time. Nobel Prize winners Arno Penzias and Bob Wilson are the people who have peered farthest into the past. They discovered the cosmic microwave background radiation, made up of microwave photons impinging on us from all directions in the sky, which is left over from the early hot infancy of the universe. These photons come to us directly from 13 billion years ago when the universe was only 300,000 years old. Our telescopes are, in a sense, time machines, allowing astronomers to sample how the universe looked at different epochs. When astronomers observe galaxies in the process of forming, it's as if a paleontologist could actually observe dinosaurs walking around. A supernova flaring up in a distant galaxy can make the evening news today, when its light reaches us, even though the event happened very long ago.

But you might also like to see past events on Earth. Even that is possible. Do you want to see yourself in the past? Stand 5 feet in front of a mirror. The image you see of yourself is not of you now but of you 10 nanoseconds ago. Traveling at 1 foot per nanosecond, light takes 5 nanoseconds to go from your body to the mirror and another 5 nanoseconds to return. So when you look in a mirror, you are seeing a slightly younger version of yourself. Using visible light, what is the farthest stretch back in time that we can observe Earth? The Apollo astronauts left some corner reflectors on the Moon. A corner reflector consists of three mirrors mounted at right angles to one another, like the floor and two side walls in the corner of a room. Bounce light into a corner reflector, and it reflects off one side and then the other and then off the floor to return exactly in the direction of the sender. (Tiny corner reflectors are used to make bicycle reflectors. They reflect headlight beams back in exactly the direction from which they came.) So scientists on Earth are now able to bounce laser beams off the corner reflectors on the Moon and have them return to Earth. The Moon is, on average, 240,000 miles away, or 1.3 light-seconds, so the round trip takes 2.6 seconds. When these scientists observe the return of the laser signal in their telescopes, they are observing an event, the sending of the laser pulse, that happened on Earth 2.6 seconds in the past. They are seeing Earth's past.

Even though we can't "see" radio waves, they too have allowed us contact with the past. The Goldstone radio telescope in California has bounced a radar signal off Saturn's rings. The round-trip time for the signal to travel from and back to Earth was 2.4 hours. So when the signal returned, the astronomers were really detecting its emission from Earth 2.4 hours earlier.

Suppose you wanted to observe Earth as it appeared a year in the past. Just put up a big corner reflector a half light-year away and look at it with a big telescope. Spy satellites 200 miles up can see license plates on cars on the ground. From 200 miles, a 6-foot-diameter telescope can resolve objects 3 inches across, which is the best resolution possible from space due to variable refraction in Earth's atmosphere. With such a telescope, from 200 miles in space you could pick out a rock star at a stadium concert. Make the telescope 10 times larger in diameter, and you could view the same scene with equal clarity from a distance 10 times farther away. The telescope will catch photons from the event at an equal rate, and you will have an equally clear view. Now, say that at an appropriate spot in the solar system you build an enormous telescope with a diameter 40 times as large as the Sun; point it at a similarly large corner reflector half a light-year away. You could then get a similarly good view of a rock concert on Earth that took place 1 year in the past. No doubt this would be an expensive project—at least 10^{31} dollars would be a guess, based on scaling up from the cost of the Hubble Space Telescope.

Reflectors already exist in space that theoretically could return photons to us from the past on Earth: black holes. Light entering a black hole never comes back out because of the enormous gravitational pull, but light traveling just outside a black hole can be bent by 180 degrees and return to Earth. The black hole Cygnus X-1, which probably weighs over 7 times as much as the Sun, is 8,000 light-years away. In principle, a photon emitted from Earth in 14,000 B.C.E. could have traveled out to that black hole, whipped around it, executing a U-turn, and headed back to Earth for arrival in the year 2000. This would provide a view of the world of 14,000 B.C.E. Unfortunately, the black hole is very small, so the fraction of all the photons emitted by Earth that come close to the black hole is tiny, and the fraction of these that actually return to Earth is also extremely tiny. Doing the calculation, one finds it likely that not even a single photon emitted from Earth has ever been returned to Earth by this black hole during their mutual history.

Another possibility for seeing our own past, suggested by the Russian physicist Andrei Sakharov, is based on the idea that the universe may be taped together in a peculiar way. For example, a flat sheet of paper obeys the tenets of Euclidean geometry, but you can roll it up and tape the left and right edges together to create a cylinder. If you were a Flatlander living on such a cylinder, you might think you were still living on a flat plane because triangles would still have a sum of angles that is

180 degrees. But if you walked around the cylinder's circumference, without changing direction, you would return to the place where you started. It's like a video game in which a spaceship goes off the left side of the screen, only to reappear instantly on the right side of the screen. The universe may be a 3-D version of this phenomenon—a giant box set up so that if you went out at the top, you would come in at the bottom; if you exited on the left, you would reenter from the right; and if you went out at the front, you would come in at the back. Light traveling from your galaxy out the front comes in the back and continues traveling forward to reach home again—your galaxy—after having made a complete round trip of the universe. In such a universe, light would circulate around and around in three dimensions, presenting many images of your own galaxy. These multiple images of your galaxy would be placed at points in a lattice (like the fish in Escher's *Depth*, Figure 8). It would look to you as if you lived in a vast universe that had many copies of your box universe stacked in three dimensions, like boxes in a giant warehouse. The nearest image of your galaxy would be at a distance equal to the distance across the shortest dimension of the box.

In 1980, I investigated such models of the universe, setting some limits on how far away the nearest image of our galaxy could be. Recent observations have improved on these limits. It appears now that if the universe is connected together in this funny way, the nearest image of our galaxy is likely to be at least 5 billion light-years away. If that is true, and if we could identify our galaxy among the billions out there, we could then see it at an epoch more than 5 billion years ago, before Earth was formed.

Neil Cornish of Montana State University, Glenn Starkman of Case Western Reserve University, and my Princeton colleague David Spergel have recently pointed out that this possibility can be tested with observations of the cosmic microwave

Figure 8. *Depth* (1955), by M. C. Escher.
Multiple images of a single fish are visible in a box universe that is "taped
together" top to bottom, left to right, and front to back.

background. The cosmic microwave background photons that we can observe come from a spherical shell with a radius of 13 billion light-years—that's as far out as we can see today. If the universe is actually a box having dimensions smaller than this, the 13-billion-light-year radius won't fit in the box so it exits the top of the box and reenters from the bottom, allowing the sphere to intersect itself. Spheres always intersect spheres in circles; in this case, the microwave background sphere reenters the box to intersect itself in pairs of circles. This means that the map of fluctuations in the microwave sky should include some pairs of identical circles. Such a pattern should be instantly recognizable, statistically, in a detailed all-sky microwave background map, as will be obtained by the Microwave Anisotropy Probe (MAP) satellite. Finding such identical circles in the microwave background sky would even tell us where to look for the nearest image of our galaxy. Just find the largest pair of identical circles in the microwave sky, and then look toward the center of one of these circles. If the nearest image of our galaxy is less than 13 billion light-years away, we could see it. It must be said that these taped-together cosmologies are not the simplest ones, so no one would be surprised if no matching circles in the microwave background turned up; finding them, however, would be very exciting. Then we might have the opportunity to see our own galaxy in the distant past—and all the great telescopes in the world would turn in that direction.

CURVED SPACETIME OPENS THE POSSIBILITY OF TRAVEL TO THE PAST

Suppose instead of just wanting to see the past, you actually wanted to go there. According to the theory of special relativity, as you move faster and faster and approach the speed of light, your clocks will slow down. If you could reach the speed of light, your clocks would stop. And if you could go even faster than the speed of light, then in principle you could go back in time—just like the "young lady called Bright."

Unfortunately, you can't go faster than the speed of light—special relativity demonstrates that, for your spaceship, it is the universe's ultimate speed limit. But according to Einstein's theory of gravity—known as *general relativity*—under certain conditions, spacetime can curve in ways that permit shortcuts through spacetime, allowing you to beat a light beam and journey back into the past.

For example, Kip Thorne of Caltech and his associates have proposed the idea of taking a shortcut back in time by traveling quickly through a wormhole—a theoretical tunnel that cuts straight across an area in which space curves. If you could take such a shortcut, you could get to a destination ahead of a light beam traveling across curved space. In that case, when you arrived, if you were to look back at your point of departure across curved space, you would see yourself preparing to leave. In fact, if you were clever enough, you might even be able to get back in time to see yourself off. General relativity allows solutions sufficiently twisted so that you could leave on a journey, come back to the place and time where you started, and shake hands with yourself as you left, a scenario I mentioned in Chapter 1.

In a sense, we are all time travelers—going toward the future at the rate of one second per second. Spacetime can be visualized as a piece of paper with time as the vertical direction and space as the horizontal direction; your world line can be shown as a straight line proceeding from the bottom to the top, always going toward the future (see Figure 9). But Einstein's theory of gravity shows that spacetime may bend. Suppose you bend the top (future) of this piece of paper around and tape it to the bottom (past), making a cylinder (Figure 9). Then your vertical world line could return to where it started by circling the cylinder, even though locally it would always seem to be traveling

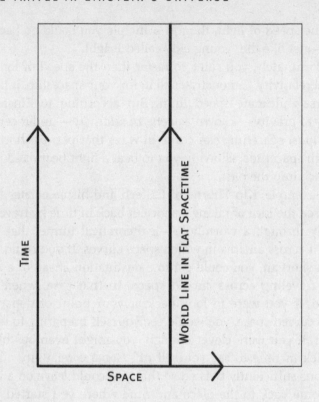

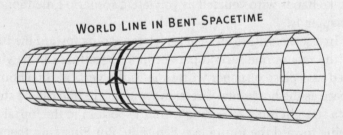

WORLD LINE IN BENT SPACETIME

Spacetime is bent into a cylinder to make a closed timelike curve.

Figure 9. World Lines in Flat and Bent Spacetime

forward in time. Your world line would complete what is called a *closed timelike curve*. In the same way, on Earth's curved surface, Magellan's crew left Europe and, traveling steadily west, eventually sailed completely around Earth to return to Europe, where the voyage began. This never could have happened if Earth were flat. Because spacetime may be curved, a time traveler may find himself revisiting an event in his own past even though, from his perspective, he has been traveling toward the future all the time.

Why Is Spacetime Curved?

A famous (perhaps apocryphal) story about Einstein describes one occasion when he fell into conversation with a man at the Institute for Advanced Study at Princeton. During their chat, the man suddenly pulled a little book from his coat pocket and jotted something down. Einstein asked, "What is that?" "Oh," the man answered, "it's a notebook I keep, so that any time I have a good idea I can write it down before I forget it." "I never needed one of those," Einstein replied. "I only had three good ideas."

One of them occurred to him in 1907—what he would later call the "happiest" idea of his life. Einstein noted that an observer on Earth and an observer on an accelerating spaceship in interstellar space would have the same sensations. Follow this chain of thought to see why. Galileo had shown that an observer dropping two balls of different mass on Earth sees them hit the floor at the same time. If an observer in an accelerating rocket in interstellar space performed the same experiment, dropping two balls of different mass, they would float motionless in space—but, since the rocket was firing, the floor of the spaceship would simply come up and hit both of them at once. Both observers thus should see the same thing. In one case, it is the result of gravity; in the other case, it is caused by an accelerating floor with no gravity involved. But then Einstein proposed something very bold—if the two situations looked the same, they must *be* the same. Gravity was nothing more than an accelerated frame-of-reference. Likewise, Einstein noted that if you get in an elevator on Earth and cut the cable, you and everything in the elevator will fall toward Earth at the same rate. (Galileo again—objects of different mass all fall at the same rate.) So, how do things look to you in the falling elevator? Any object you drop will float weightless in the elevator —because you, the object, and the elevator are all falling at the same rate together. This is exactly what you would see if you were in a spaceship floating in interstellar space. All the objects in the spaceship, including you, would be weightless. If you want to experience weightlessness just like an astronaut, all you have to do is get in an elevator and cut the cable. (This works, of course, only until the elevator hits bottom.)

Einstein's assertion that gravity and acceleration are the same—which he called the *equivalence principle*—was influenced, no doubt, by his previous success in equating the situation of a stationary magnet and a moving charge with that of a stationary charge and a moving magnet. But if gravity and accelerated motion were the same, then gravity was nothing but accelerated motion. Earth's surface was simply accelerating upward. This explained why a heavy ball and a light ball, when dropped, hit the floor at the same time. When the balls are released, they just float there—weightless. The floor (Earth) simply comes up and hits them. What a remarkably fresh way of looking at things!

Still one must ask how Earth's surface could be accelerating upward (away from Earth's center) if Earth itself is not getting bigger and bigger with time like a balloon. The only way the assertion could make sense is by considering spacetime to be curved.

Einstein proposed that mass and energy cause spacetime to

curve. It took him 8 years of hard work to derive the equations governing this. He had to learn the abstruse geometry of curved higher dimensional spaces. He had to learn about the Riemannian curvature tensor—a mathematical monster with 256 components telling how spacetime could be curved. This was very difficult mathematics, and Einstein ran upon many false leads.

But he didn't give up because he had great faith in the idea. He also had some competition. In the summer of 1915, when he gave a talk describing his idea and his mathematical difficulties, the great German mathematician David Hilbert was in the audience. Hilbert set about trying to solve the problem himself. He found the correct equations by using a sophisticated mathematical technique Einstein was not using. Nearly simultaneously, Einstein arrived at the same equations himself. There has been some dispute among historians of science as to who submitted the equations in their final form first, a dispute now apparently resolved in Einstein's favor. The terms in the equations were complicated mathematical objects called tensors, but the equations themselves were beautiful and simple. If you want to know how Einstein's equations look, here they are—ten independent equations rolled into one: $R_{\mu\nu} - \frac{1}{2}g_{\mu\nu}R = 8\pi\, T_{\mu\nu}$. The left side of the equation tells how spacetime is curved at a particular place, and the right side of the equation refers to the mass-energy density, pressure, stress, momentum density, and energy flux at that place, all of which cause spacetime to curve. Einstein had shown that mass could be converted into energy and vice versa, but the total amount of mass (times c^2) plus the amount of energy was a constant quantity. These equations of general relativity implied that the law of mass-energy conservation (you don't get mass or energy out of nothing) was automatically valid in every tiny region of spacetime. Furthermore, the equations approximate Newton's laws for circumstances in which spacetime is nearly flat.

Einstein's derivation of his equations for gravity, having only Newton's theory of gravity to go on, is as remarkable as if Maxwell had derived all the equations of electromagnetism knowing only the laws of static electricity and nothing else. Maxwell had many more hints: he knew about magnetic fields and even had some equations involving them. Einstein had no such hints, and the math was much more difficult. He kept plugging away, trying different ideas until he got it right. Einstein said of his travails, "But the years of anxious searching in the dark, with their intense longing, their alternations of confidence and exhaustion, and the final emergence into the light— only those who have experienced it can understand that." (One of them would be Princeton mathematician Andrew Wiles, who finally proved Fermat's last theorem—a longstanding unsolved mathematical challenge—after seven years of effort.)

When Einstein finally had the equations right, the theory made some remarkable predictions. In Einstein's theory, planets would travel along geodesics—the straightest trajectories in curved spacetime. To grasp this idea, think of a jetliner traveling on a great-circle route (a geodesic) from New York to Tokyo. It will always travel straight ahead—the pilot steers neither left nor right yet the route is curved. Find this great-circle route by stretching a string between the two cities on a globe. The string should be taut, as straight as it can be, and yet that path should pass north of Alaska. Try it. If you track that plane's trajectory on a classroom Mercator map of Earth, it will look curved. Similarly, Earth's world line appears as a helix in spacetime, circling the world line of the Sun (return to Figure 1). Yet Earth's world line is as straight as can be in the curved geometry formed by the mass of the Sun warping spacetime around itself.

Einstein's theory explained exactly a well-known peculiarity observed in the orbit of Mercury long famous for its disagreement with Newton's theory of gravity. The long axis of Mercury's elliptical orbit around the Sun was slowly shifting direction (precessing) by an excess amount of 43 seconds of arc per century (a second of arc is one 3,600th of a degree). But when Einstein calculated the geodesic corresponding to Mercury's orbit, he found an extra twist of exactly 43 seconds of arc per century. Eureka! Einstein was so excited doing this calculation that he said he had heart palpitations.

Einstein made another prediction—that light beams bend when traveling near the Sun. And this effect could be checked. Just take a picture of the stars in the sky near the Sun during a total solar eclipse, when you can see stars near the Sun. Compare this photo with one taken 6 months earlier when the Sun was on the side of the sky opposite these stars. The two pictures should look slightly different because of the stronger bending of the stars' light beams passing near the Sun during the eclipse. Einstein's theory predicted a deflection of 1.75 seconds of arc for light beams passing near the edge of the Sun, twice the amount of deflection predicted by Newton's theory if photons (like high-speed bullets) were attracted by the Sun just like planets are. (A no-deflection result could still vindicate Newton because photons would travel straight if they were not attracted by gravity. Einstein's theory required a deflection because in his theory the photons were already traveling in the straightest possible trajectories allowed in the curved geometry.) Since a total solar eclipse was expected on May 29, 1919, here was the opportunity for a real test—with predictions made in advance. If light passing near the Sun was deflected by 1.75 seconds of arc, Einstein would be right; if a deflection of either zero or 0.875 seconds of arc was observed, Newton would win.

Two expeditions were mounted to take measurements from two different locations where the eclipse would be visible: Sobral, Brazil, and Principe Island off the coast of Africa. As recounted by Abraham Pais, Einstein's biographer, the results were announced at the November 6, 1919, combined meeting of Britain's Royal Society and its Royal Astronomical Society. The measurement from the Sobral eclipse expedition was 1.98±0.30 seconds of arc, and the measurement from the Principe expedition was 1.61±0.30 seconds of arc. Both results agreed with Einstein's value of 1.75 seconds of arc, to within the allowance for observational uncertainties (±0.30 seconds of arc), and both disagreed with the Newtonian values. Nobel Prize winner J. J. Thomson, the discoverer of the electron, chaired the meeting and, after hearing these results, pronounced, "This is the most important result obtained in connection with the theory of gravitation since Newton's day, and it is fitting that it should be announced at a meeting of the Society so closely connected with him. . . . The result [is] one of the highest achievements of human thought." The next day the *London Times* carried the story with the headline "Revolution in Science." The *New York Times* picked up the story two days later. The world was ready to embrace Einstein's vision.

GÖDEL'S UNIVERSE

Ever since Einstein announced his equations of gravitation in 1915, people have been exploring "solutions" to them. In the language of physicists, such a solution gives both a mathematical description of the geometry involved, how the spacetime would look, and the distribution of mass and energy required to produce it. Many of these solutions have remarkable properties. One of the most amazing was found in 1949 by Einstein's brilliant colleague at the Institute for Advanced Study at Princeton, mathematician Kurt Gödel. The solution allowed time travel to the past.

Gödel's remarkable solution to Einstein's equations was a universe that was neither expanding nor contracting but instead rotating. Now put aside thinking about the universe for a moment, and consider yourself. Your inner ear tells you whether you are spinning or not—if you are rotating rapidly, you will get dizzy. In that case, the fluid in your inner ear is drawn outward within its semicircular canals, giving your brain conflicting ideas of the direction of up. Your brain says "uh-oh," and you become dizzy. Alternately, you can tell that the room around you is not spinning rapidly by noting that your body is at rest with respect to the room and you are not feeling dizzy. If someone were to kidnap you and put you in a funhouse room that was on a rapidly spinning merry-go-round, you would know that the room was rotating because if you held your body in a fixed position relative to the room, you would become dizzy. The only way not to become dizzy would be to keep turning in the direction opposite to the room's rotation, counteracting its spin. (In principle, if your inner ear were much more sensitive, you could use this technique to tell that Earth is rotating—but it rotates too slowly for you to detect.)

Back to Gödel's universe. In that universe, a nondizzy, and therefore non-rotating, observer would see the whole universe spinning around her. From this, she could conclude that the universe was rotating. Furthermore, the distances between galaxies in Gödel's universe do not change with time; they are like fixed entrées on a giant rotating lazy Susan. A nondizzy observer might then suppose that galaxies far enough away from her would be traveling faster than light, as they circled her in giant circles. This does not conflict with the results of special relativity because that just says that the relative velocity of galaxies as they *cross paths* with each other cannot exceed the velocity of light; the galaxies in Gödel's universe never cross paths but simply stay at fixed distances from one another. (You could equally view Gödel's universe as static and non-rotating, as long as self-confessed "nondizzy observers" would be spinning like whirling dervishes with respect to the universe as a whole.)

A photon sent out in Gödel's universe would try to go in a

straight line, but given the rotating universe, would actually execute a wide turn like a boomerang. Gödel's universe has an even more curious property. If you set out from your galaxy and made a short trip, you would come back after you left. But if you went on a long enough journey at a velocity near to but lower than the speed of light, you could actually return home at the time you started or even before. Since light follows looping, boomerang-like trajectories in Gödel's universe, you could fire your rocket continuously in such a way that you could cut straight across the sweeping boomerang path and beat the light beam. Take advantage of this on a long enough journey and, like Miss Bright, you could return home the previous night. Gödel was smart enough not only to understand Einstein's theory but to spin it in a new direction—time travel.

Yet our observations tell us that we apparently do not live in the universe proposed by Gödel. We observe that galaxies are moving away from each other—the universe is expanding. With all the orbits of the planets, asteroids, and comets, the solar system constitutes a giant gyroscope, and we can determine that the distant galaxies are not rotating relative to it. Also, if the universe had a significant amount of rotation, the temperature of the cosmic microwave background would vary in a systematic way over the sky—something we don't observe. Nevertheless, the Gödel solution is very important, for it showed that time travel to the past is possible in principle, with Einstein's theory of gravity. If there is one solution that has this property, there can be others.

COSMIC STRINGS

Let's look at another exact solution to Einstein's equations— one describing the geometry around a *cosmic string*. This term refers to thin strands of high-density material left over from the early universe, which are predicted in about half the proposed theories attempting to unify the different forces in the universe (thereby explaining all the laws of physics). Given these proposals that cosmic strings are likely to exist, we wouldn't be too surprised to discover them. But finding them would surely be exciting! One of the major candidates for a theory-of-everything is superstring theory, which, as I mentioned in Chapter 2, suggests that even elementary particles such as electrons are really tiny loops of string. Superstrings are theoretically of zero width and form microscopic closed loops, whereas cosmic strings have a tiny (nonzero) width and may be millions of light-years long, or longer.

Cosmic strings have no ends, and so, in an infinite universe, are either infinite in length or exist in closed loops. Think of infinite strands of spaghetti, or Spaghetti-o's. Physicists who predict the existence of cosmic strings expect both varieties but anticipate that most of their mass will take the form of infinitely long strings. Scientists figure that cosmic strings should have a width narrower than an atomic nucleus and a mass of about 10 million billion tons per centimeter. Strings are also under tension, like stretched rubber bands, which causes infinite strings to straighten out with time and whip around at velocities that should typically be over half the speed of light.

Since cosmic strings are so massive, they should warp the spacetime around them. But how? Alex Vilenkin of Tufts University found an approximate solution to Einstein's equations for a straight, infinitely long cosmic string, valid as long as the geometry of spacetime around the string is approximately flat. According to Vilenkin's solution, slices through the string would look like cones rather than sheets of paper. This gave me a big clue as to the exact solution and its appearance. Previously, in 1984, with my student Mark Alpert, I had studied how general relativity would work in Flatland, with two dimensions of space. We found that for a massive body in Flatland, there was an exact solution whose exterior geometry was shaped like a cone. (Two other groups of physicists—Stanley Deser, Roman Jackiw, and Gerard 't Hooft, as well as Steven Giddings, J. Abbott, and Karel Kuchař—had similar findings, publishing the same year. A Polish physicist, A. Staruszkiewicz, it turned out, had explored the topic in a preliminary way 20 years earlier.) I hypothesized that adding a third vertical dimension to our solution in Flatland could provide an exact solution for a cosmic string.

I plugged my guess about the shape of the spacetime into the left side of Einstein's equations to discover whether they would give me the correct string density and tension on the right side. This required me to solve the equations both within and outside of the string—Einstein's equations must be satisfied everywhere. They worked; I now had an exact solution.

William Hiscock of Montana State University found the same solution independently. I published in the *Astrophysical Journal*, Hiscock in the *Physical Review*, and today we are given joint credit for this solution. (Later, French physicist Bernard Linet added some details and American physicist David Garfinkle contributed some additional particle physics.) Linet, looking back through physics literature, noticed that this geometry had been proposed in 1959 by L. Marder, of the Department of Mathematics at the University of Exeter, who saw it as a mathematical solution to Einstein's equations without realizing it could apply to cosmic strings. In fact, Marder's work appeared before cosmic strings were even suggested, and his result was almost forgotten. This shows that one should pay attention to beautifully shaped spacetimes—they often turn out to be physically relevant.

Here's how to visualize our solution for the geometry of the spacetime around a straight, infinitely long cosmic string. Sup-

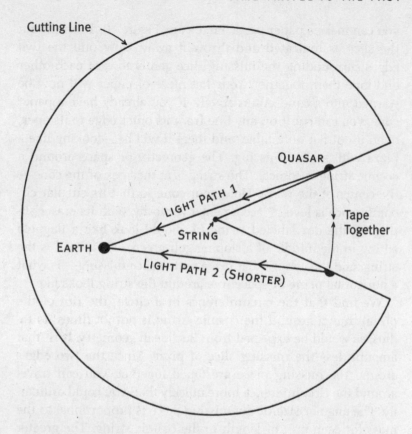

Figure 10. Space Around a Cosmic String

pose the string is vertical. In that case, one might naively expect that a horizontal plane cutting through the string would look like a flat piece of paper with the string appearing as a dot in the middle of the page. But we found that such a plane looks like a pizza with a slice missing. When I explain this to my relativity classes at Princeton, I usually order pizza for the entire class to illustrate the point. But if you don't have a real pizza, you can make a paper pizza. First, copy Figure 10. Then cut out the slice as indicated and throw it away. Now pull the two edges surrounding the missing slice gently toward each other and tape them together. Your flat piece of paper will now be warped into a cone. Alternatively, if you already have a paper cone, you can cut it on any line from its outer edge to its apex, open it out flat on a table, and there it will be—looking like a pizza with a slice missing. The geometry of space around a cosmic string is conical. The string is at the apex of the cone— the center of the pizza. Hold your cone so that its circular circumference is level. Place a pencil vertically, with its eraser sitting on the dot labeled *string* (it should look like a flagpole sitting in the middle of a sloping golf green). The pencil is the string, and the cone—that pizza with a slice missing—is what a horizontal plane of spacetime around the string looks like.

We find that the circumference of a circle (the rim of the pizza) drawn around the cosmic string is not 2π times its radius, as would be expected from Euclidean geometry. It is that amount, less the missing slice of pizza. Since the two edges around the missing piece are taped together, you can travel around the circumference more quickly than you could ordinarily. The angular size of the missing piece is proportional to the mass for each unit of length in the cosmic string. The greater that amount, the larger the missing piece and the steeper the slope of the cone. Given the prediction that a cosmic string has a mass of 10 million billion tons per centimeter, the missing slice would subtend an angle of 3.8 seconds of arc. That's a very thin slice; just one 340,000th of the pizza would be missing. Yet though this distortion of space isn't big, it is measurable.

Suppose a cosmic string lies about halfway between us and a distant quasar. (Quasars can be seen up to 12 billion light-years away.) Cut through the tape on the cone so that the "quasar dot" shown on the paper is split. Part of the dot then appears on each edge surrounding the missing "pizza slice." Notice the straight line extending from each of the quasar images to Earth (as shown in Figure 10). These are paths that light will take in the conical spacetime. Each of the two images of the quasar is connected to Earth by the shortest, most direct path. Light beams will travel on these two paths straight to Earth, which means that light from the quasar will be arriving on Earth from two slightly different directions. Consequently, an observer on Earth will see two images of the distant quasar, one on each side of the cosmic string. Image 1 will appear to the left of the string and image 2 to the right. Those two images will lie in the directions of the two straight paths (labeled 1 and 2 in the figure). The angular separation of the two images in the sky as seen from Earth will be about half the angular width of the missing "slice," or about 1.9 seconds of arc. You've just grasped the principle of gravitational lensing—light beams that are bent by the geometry of spacetime.

This phenomenon allows us to search for cosmic strings. A string, of course, would be too thin to see, but identifying a string might be undertaken by looking for double images of background quasars. Pairs of quasars with identical spectra and equal brightness should be laid out in the sky like pairs of buttons in a double-breasted suit; threading between the pairs should be a cosmic string. Final confirmation would come from radio astronomy. Radio telescopes can map the sky at radio wavelengths. Since we expect the string to be moving rapidly, the cosmic microwave background photons on each side of it should undergo either slight redshifting or blueshifting as they whip around opposite sides of the moving string. An accurate radio map would show the string snaking across the sky, as a line dividing a slightly hotter region from a colder one. Such a discovery of a cosmic string would be of enormous importance. Not only would it give us new clues about the very early universe, but it would certainly boost the hopes for a theory-of-everything.

An important point: when we see two images of a distant quasar, the distances to the two images along the two paths can differ slightly. In your copy of Figure 10, for example, if you take a ruler, you will see that the two straight lines connecting Earth to the two quasar images differ in length. The bottom path is shorter. Since light always travels at 300,000 kilometers per second, if one path is shorter than the other, a light signal from the quasar coming along the shorter path arrives sooner.

Similar effects occur when light is bent while traveling around opposite sides of a massive galaxy. A group led by my Princeton colleagues Ed Turner, Tomislav Kundić, and Wes Colley, in which I participated, has observed the gravitationally lensed quasar 0957, which has two images, A and B, on opposite sides of such a galaxy. This quasar varies noticeably in brightness with time. We observed a sharp drop in the brightness of image A, and given the lensing geometry, we predicted that this should be followed by a similar drop in the brightness of image B, whose light we expected to reach us after a slightly longer time. We published our prediction and continued to observe; 417 days later there was an identical brightness drop in image B. This time difference was a tiny fraction of the total travel time of approximately 8.9 billion years.

This shows that you can beat a light beam in a race. Light beam A beat light beam B by taking a shortcut through spacetime. A spaceship traveling at 99.9999999999 percent of the speed of light along path A would have done just a little worse at the race, still beating a photon traveling along path B by 414 days.

If cosmic strings exist, you could travel in a spaceship and outrun a light beam by taking the shorter of two paths around a cosmic string. The door to time travel to the past begins to crack open.

Cosmic Strings and Time Travel to the Past

My idea for a time machine to visit the past is based on an exact solution to Einstein's equations, which I published in 1991. Here is the scenario. First, imagine placing two infinitely long, straight cosmic strings parallel to each other like two flagpoles. Interestingly, they do not attract each other gravitationally—they just sit there, motionless. That's because although the strings have a great mass density inside them, they are also under tension, like taut rubber bands. This tension, which tends to pull the string together, is produced by a negative pressure, or suction, within the string. The negative, repulsive gravitational effect of this negative pressure exactly offsets the gravitational attraction of the mass density in the string. Thus, if we set two cosmic strings near each other, at rest, they would stay in the same position.

To visualize the cross section through the spacetime perpendicular to the two strings, copy Figure 11 and cut the pattern out as indicated. Your paper has two dots representing the two cosmic strings, but, as in the pizza example, missing wedges extend from each string. Place two pencils vertically with their erasers sitting on the dots labeled String 1 and String 2. The two "strings" are like two flagpoles, standing straight up, with the figure showing a horizontal cross section of the surrounding geometry. Now tape the two sides of the upper V together and the two sides of the upside-down V together to model the spacetime. It resembles a paper boat.

Next, imagine two planets, A and B, to the right and left of the strings. Suppose you lived on planet A and wanted to visit planet B. You could do this by traveling directly to planet B along path 2, between the two cosmic strings. That is a geodesic path—a straight path you could navigate between planets A and B. But there is another straight path from A to B, path 1, which goes around the top of cosmic string 1. If you measure carefully, you will see that the total distance from planet A to B

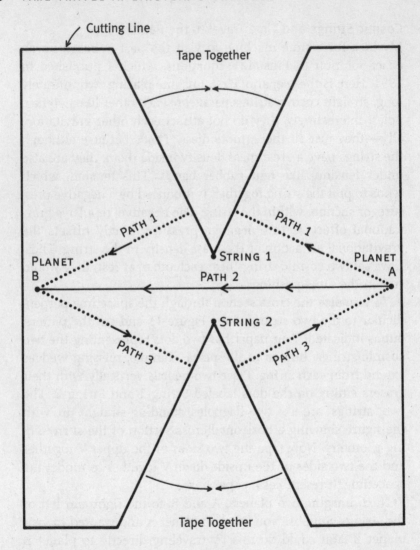

Figure 11. Space Around Two Cosmic Strings

along path 1 is slightly shorter than along path 2, because of the missing wedge. Path 1 is a shortcut from planet A to B. Send a light beam from A to B along path 2, and you can get in your rocket ship, travel at 99.999999 percent of the speed of light, and beat the beam by traveling along path 1, around string 1. When you arrive at planet B, the light beam showing your departure will not have arrived yet. When you look back at your home planet along path 2, you will therefore see yourself on planet A, getting ready to depart.

Interesting. Perhaps, if you are clever enough, you may still have time to return and see yourself off. In fact, there is an observer (let's call him Cosmo) traveling rapidly in a rocket ship along path 2 from planet A to B who will think that your departure from A and your arrival at B are simultaneous events. Why? Because you have beaten the light beam traveling along path 2, your departure and arrival are two events separated along path 2 by more light-years in space than years in time. Since this results in a spacelike separation, Cosmo can see those events as having a separation in space but no separation in time.

I noticed I could exactly divide the spacetime shown in Figure 11 by making a sharp cut along path 2 (imagine bringing a meat cleaver straight down on path 2). Do so on your model, and you will separate the spacetime into the top half, containing string 1, and the bottom half, containing string 2. Because the whole static spacetime could be cut out of a flat spacetime by just excising two wedges, the boundary (the cleaver cut) between the two halves of the spacetime is also exactly flat. The boundary has no intrinsic curvature and is not bent. In other words, the top half of Figure 11 can slide toward the right at high velocity (but less than the velocity of light), and the bottom half of Figure 11 can slide toward the opposite direction at an equally high velocity; the two boundary surfaces still fit together perfectly as they slide past each other.

Using this thinking, I produced a geometry in which string 1 moves rapidly to the right, string 2 moves rapidly to the left, and the two halves of the spacetime fit together perfectly. This exactly solves Einstein's equations for both halves and along the boundary between them. To ensure that our friend Cosmo will not be split in two by this process, let's imagine putting him slightly in the upper half of the diagram so that he, along with string 1, is dragged toward the right. In fact, we will be moving the top half of the diagram at exactly the speed required to compensate for Cosmo's initial velocity and bring him to rest.

Now think of Cosmo sitting at rest midway between planets A and B on path 2. He sees string 1 moving at nearly the speed of light toward the right and string 2 moving at nearly the speed of light toward the left. If you then travel between planets by taking the shortcut around the back of string 1, Cosmo will see you depart from planet A at noon and arrive at planet B, also at noon. You pull off that trick by traveling against the motion of string 1 along path 1 (see Figure 12). Since string 2 goes in the opposite direction, you can accomplish that trick a second time by going against string 2's motion, on the return trip to planet A by path 3. You can thus leave planet B at noon and return to planet A at noon, according to Cosmo's observations. Since Cosmo thinks your departure from planet A and your arrival back there occur at the same place (planet A) and time (noon), they are one event.

How would the trip look to you? It is like the story described in Chapter 1. When you arrive at the planet A spaceport at noon, you will see there a slightly older version of yourself who will greet you by shaking hands and saying "Hello, I've been around the strings once!" You'll reply, "Really?" You will then climb into your rocket and fly to planet B by whipping around the rapidly approaching string 1 along path 1. Then you will

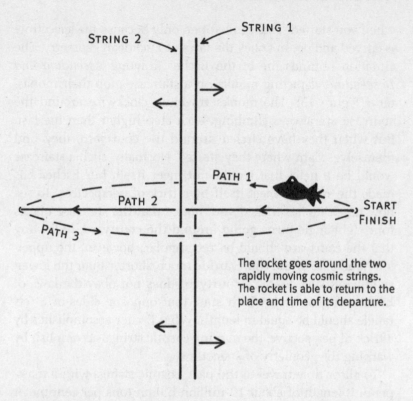

The rocket goes around the two rapidly moving cosmic strings. The rocket is able to return to the place and time of its departure.

Figure 12. Traveling Back to a Past Event

return to planet A along path 3 by whipping around the rapidly approaching string 2. Arriving back at planet A at noon, you will see your slightly younger self off, shaking hands and saying, "Hello, I've been around the strings once!" You will have accomplished time travel to an event in your past.

The moving cosmic string solution is sufficiently twisted to allow you to travel counterclockwise around the two moving cosmic strings, always toward the future, and still arrive back when you started. This can happen only because the spacetime is curved and doesn't obey the laws of Euclidean geometry. The situation reminds me of the Escher drawing *Ascending and Descending*, depicting monks on a staircase atop their monastery (Figure 13). The monks traveling clockwise around the staircase are always climbing, each step higher than the last. But when they have circled around the courtyard, they find themselves right where they started. Normally, such a staircase would be a helix that would not meet itself, but Escher has made the staircase meet itself by a trick of perspective. To see how much he has tricked you, notice that the staircase makes four right-angle turns going around the courtyard, indicating that the courtyard should be rectangular; however, the upper-left-hand side of the courtyard is much shorter than the lower-right-hand side. Escher's courtyard does not obey the laws of Euclidean geometry, which state that opposite sides of a rectangle should be equal in length. What Escher accomplishes by a trick of perspective, the moving cosmic strings accomplish by warping the geometry of spacetime.

To allow time travel to the past, cosmic strings with a mass-per-unit length of about 10 million billion tons per centimeter must each move in opposite directions at speeds of at least 99.999999996 percent of the speed of light. We have observed high-energy protons in the universe moving at least this fast, so such speeds are possible.

When I found this solution, I was quite excited. The solution used only positive-density matter, moving at speeds slower than the speed of light. By contrast, wormhole solutions require more exotic negative-energy-density material (stuff that weighs less than nothing). I checked the solution a number of times, wrote it up, and sent it into the *Physical Review Letters*, one of the world's premier journals for fast publication. I told no one and waited for the replies of the reviewers. Two reports came back, approving my work, suggesting only a couple of minor

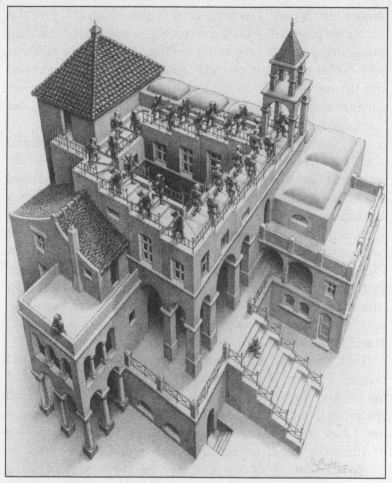

Figure 13. *Ascending and Descending* (1960), by M. C. Escher.

additions. Finally, the paper appeared, on March 4, 1991. I went to the Institute for Advanced Study, Einstein's old place of work, to make a copy of the journal article, since the institute somehow receives issues a day or two before Princeton University's physics library does. I took the copy to show John Wheeler, the Princeton physicist who invented the term *black hole*. Kip Thorne happened to be coming to Princeton to give a talk that week on his time-travel research using wormholes, so I showed him the copy as well. In the movies, scientists always explain things to each other by scrawling equations on blackboards. But I explained my solution by showing paper cutouts.

Later that day Thorne mentioned my new result at the end of his talk. In the hallways and coffee lounges of scientific departments, ideas and research papers are continually discussed and debated. Although my paper was widely recognized as a remarkable solution to Einstein's equations, it naturally created a stir, with some skeptics doubting whether such time travel could actually occur in our universe. Alex Vilenkin of Tufts University invited me to address the Tufts-Harvard-MIT relativity group in Boston. To my delight, the room was packed with many eminent scientists. Bill Press came over from Harvard, and Alan Guth brought his colleagues, Edward Farhi and Sean Carroll, from MIT.

(That same day, an article about my research by Michael Lemonick appeared in *Time* magazine. It included a picture of me holding up two strings and passing around them a small spaceship that my 7-year-old daughter had given me. Years earlier, I had appeared in *Newsweek* holding up one string—to illustrate the one-string solution. This explains the otherwise curious fact that there is a picture of me in *Newsweek* holding one string, and a picture of me in *Time* holding two!)

Guth and his two MIT colleagues would later find some interesting properties of my solution, including the fact that by the time your rocket returned to planet A, it would have been spun around by 360 degrees as well as acquiring a kick in velocity. Kip Thorne took news of my solution back to Caltech where a student of his, Curt Cutler, found an even more intriguing property. Cutler decided to see if every event in my spacetime could be visited twice by a time traveler. All events that a time traveler could visit twice, like your departure from planet A, belong to a geometric region of time travel. Any events that no time traveler could ever return to would belong to a no-time-travel region. Cutler found that my spacetime included both types: a region circling the strings, where time travel to the past is possible, surrounds an hourglass-shaped region of spacetime, where time travel to the past is impossible (see Figure 14). In this spacetime diagram, two dimensions of space are depicted horizontally while time is depicted as the vertical dimension. Up is toward the future. String 1 is moving to the right with time, and the world line it traces is shown as a diagonal line tilted upward and to the right. Farther and farther into the future we see string 1 farther and farther to the right. String 2, moving in the opposite direction, has a world line tilted upward and to the left. In the distant past (the bottom of the image), string 1 is to the left of string 2. They cross in the center, and in the distant future (the top of the image), string 1 is to the right of string 2. The surface separating the region of time travel from the region where time travel to the past is impossible looks like a lampshade and an upside-down lampshade, glued together. This surface is called the *Cauchy horizon*. (It's named after the nineteenth-century French mathematician Augustin-Louis Cauchy, who did some related mathematical work.) Events inside the hourglass shape cannot be visited again. Events in the region outside the hourglass, circling the strings, can be visited again by a time traveler.

The world line for you—the time traveler—is also shown. You start at the bottom right of the diagram, sitting still on planet A. Your world line ascends vertically because you are not moving in space but only moving forward in time. Then you depart and circle the strings—the horizontal circle shown.

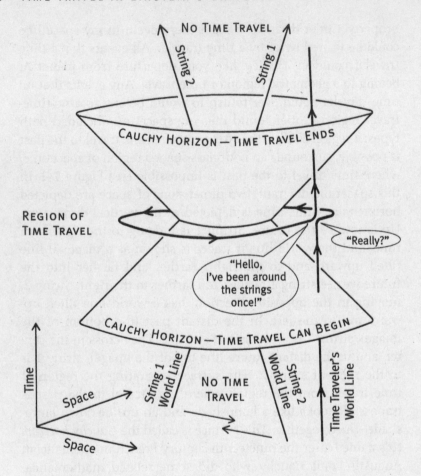

Figure 14. The Region of Time Travel Around Two Cosmic Strings

When you return to planet A, you say, "Hello, I've been around the strings once!" After that, you simply stay on the planet, and your world line continues vertically upward. The event at which you meet yourself and say hello is in the time-travel region. Interestingly, the hourglass-shaped Cauchy horizon bounds the region of time travel to both the past and future. Note that your world line starts out at the bottom in the distant past, in the region where no time travel is possible. In the distant past, the two cosmic strings are so far apart that any traveler starting on planet A at that time would take so long getting around them both that she would always arrive back home after starting. As the strings move closer and your world line pierces the Cauchy horizon and enters the region of time travel, it suddenly becomes possible to come back and shake hands with yourself. The time machine has been created, and for a while your world line remains within the region of time travel. But eventually, your world line passes back outside the upper inverted lampshade surface, and the possibilities for time travel are over for you. The time machine has been destroyed. The strings are now so far apart that by the time you circled them, you would always arrive back after you left. Time travel is possible only in the interval when the time machine is in existence.

This answers Stephen Hawking's famous question: "Why haven't we been overrun by tourists from the future?" It's simply because no one has built a time machine yet. In lay terms, if a time machine were built in the year 3000, a time traveler could perhaps use it to go from the year 3002 back to the year 3001, but she couldn't use it to go back to the year 2001, because that was before the time machine was built. Time machines such as my string solution and Kip Thorne's wormhole solution, which involve twisting spacetime, both include regions of spacetime in which time travel isn't possible. If no time machines have been built yet, we on Earth now cannot visit the past. Furthermore, all the events of which we can currently have any knowledge lie inside our past light cone, also before the region of time travel. Thus, we see no time travelers at the Kennedy assassination in 1963. Important as that event was, it, like us, lies before any time machines were created, and so no time travelers can visit it. Yet Cutler's work shows that even if observers check their own past very carefully and never find any evidence of time travelers, they shouldn't conclude that they will never encounter any time travelers in the future. At any time an observer might cross a Cauchy horizon and suddenly enter a region of time travel, where time travelers from the future may unexpectedly show up and say hello.

COSMIC STRING LOOPS AND BLACK HOLES

Suppose you wanted to construct a time machine based on cosmic strings, but you were not lucky enough to find two infinitely long cosmic strings passing each other at the requisite high speed in our universe. You might, however, find a large loop of cosmic string in space. Such a loop would be like a giant, oscillating rubber band under such great tension that it could snap shut. A supercivilization could always manipulate such a loop gravitationally by flying massive spaceships near it until it acquired the right spin and assumed the desired shape. If the original loop was arranged just so—in a slightly bent rectangular shape, like the frame outlining a reclining lawn chair —it would collapse and, as it did so, two straight sections of the loop would pass by each other at a speed high enough to create a time machine.

A collapsing loop of string large enough to allow you to circle it once and go back in time a year would have more than half the mass-energy of an entire galaxy. But a worse problem exists —such a massive string loop would become so compact as it collapses that it would be in danger of forming a black hole.

A black hole is a cosmic Hotel California: you can check in, but you can't check out. Normally, when you throw a ball up into the air, it falls back down to Earth. But throw a ball up at a speed of greater than 25,000 miles per hour, Earth's escape velocity, and it will not return. Astronauts going to the Moon had to achieve such velocity. Escape velocity is the key to understanding black holes. If you were to compress Earth's mass to a smaller size, its escape velocity would rise. Compress Earth until its circumference was less than 5.6 centimeters, and its escape velocity would become greater than the velocity of light. Since nothing can go faster than light, nothing could escape from our compressed Earth—it would have become a black hole. Under this condition, gravity would cause Earth to continue to collapse, rapidly forming a *singularity*—a point of infinite density and curvature. Actually, quantum effects might limit the singularity's density to about 5×10^{93} grams per cubic centimeter, but still its size would be smaller than an atomic nucleus. Surrounding this tiny singularity would be just curved, empty space. Enclosing the singularity would be a spherical *event horizon*. Anything that happens inside that sphere with a circumference of 5.6 centimeters would remain forever hidden from observers outside the sphere, because any light emitted inside is unable to escape. (The size of the event horizon of a black hole depends on its mass. A black hole with a mass 3 billion times larger than the Sun, such as has been observed by the Hubble Space Telescope in the nucleus of the galaxy M87, has an event horizon with a circumference of 56 billion kilometers, or about 52 light-hours.)

Suppose a professor wanted to investigate a non-rotating 3-billion-solar-mass black hole. The professor might remain safely outside the black hole, 34.2 light-days away, and send in her graduate student. As the unfortunate graduate student falls in, he radios back observations. His message: *"Things are going badly!"* The graduate student sends the word "going" just as he crosses the event horizon—nothing bad has happened so far. The graduate student takes 18 months to reach the horizon, as measured by his watch. He notices nothing unusual as he crosses it. No warning sign is posted there. But once he crosses the event horizon, he travels past a point of no return. Now, no matter how he fires his rocket, he is drawn inexorably toward the singularity at the center of the black hole. The spacetime inside the black hole is so warped that the singularity now looms in the future of our poor graduate student who can no more avoid hitting it than you can avoid next Tuesday. If, as he falls in, his feet are closer to the center of the black hole than his head, his feet will be pulled inward more strongly than his head. He will be stretched as if he were on a rack. Furthermore, since each shoulder wants to fall straight into the hole, his two shoulders will wedge together as he approaches the center—as if he were being crushed in an iron maiden as well. The tidal force stretching and crushing him will become ever stronger. As he gets closer to the singularity, the near-infinite spacetime curvature will stretch him like a piece of spaghetti, ripping his body apart. His remains then will be deposited in the singularity at the center. The black hole has now added the mass of one unlucky graduate student to its mass. As measured by the graduate student's watch, about 5.5 hours would have passed from the time he fell inside the event horizon until he was shredded and deposited in the singularity.

Meanwhile, the photons he had sent back to the professor as part of his radio message work their way outward. The word "Things," emitted well outside the black hole, is received quickly by the professor. The word "are," which is sent just outside the event horizon, may take thousands of years to climb out. The word "going," which is emitted right at the event horizon, travels outward at the speed of light, of course, but, like a kid running up a down escalator, it makes no progress. The word just stays at the event horizon, running in place. And the word "badly," sent from within the black hole just before the student's messy death, behaves like a kid running up a super-fast down escalator. Although running "upward," the signal is being drawn backward even faster and is eventually drawn into the singularity, just like the graduate student. The professor receives the message as "Things a . . . r . . . e . . ." She never finds out what happened to the student inside the black hole's event horizon. That's why it's called a horizon—you can't see past it. If the professor later follows the graduate student into the black hole, the professor will encounter the signal "going" just as she crosses the event horizon—it's still there, of course— and as she falls in, she will see it pass her at exactly the speed of light, consistent with special relativity.

Now that you're warned of the dangers of being an astrophysics graduate student doing field investigation of black holes, let's consider what this means for my string loop time machine. As noted in my original *Physical Review Letters* paper on moving cosmic strings, at the very moment when the collapsing string loop is reaching the critical velocity to allow time travel, its circumference is becoming so small that, given its mass, it is sufficiently compressed to be in danger of forming a rotating black hole. Here I had used a criterion called the "hoop conjecture," proposed by Caltech physicist Kip Thorne. Thorne had argued that if a lump of mass were compressed enough for its circumference in every direction to be smaller than the circumference of the event horizon of a black hole with that same mass, then the lump would always collapse to form a black hole itself. It's not a proof that a black hole will form in this case, but it's a good argument, and no exception to Thorne's hoop conjecture has been found so far. Because of the way the two straight segments pass each other, the string loop has some angular momentum, so it must form a rotating black hole. If a black hole forms as expected, any possible regions of time travel could well be trapped inside the black hole. Here are three possible ways this could play out.

1. One could fall into the rotating black hole and be killed (torn apart by near-infinite spacetime curvature) before being able to do any time travel to the past.

2. One could fall into the rotating black hole and travel back in time but not be able to get back out to brag to friends about it. Later the time traveler would be killed by being torn apart by near-infinite spacetime curvature.

3. One could fall into the rotating black hole, travel back in time, and later emerge into a different universe. Maybe that's better—but bragging to old friends could still never occur.

In 1999 physicists Sören Holst, from the University of Stockholm, and Hans-Jürgen Matschull, from Johann-Gutenberg University in Mainz, Germany, discovered an exact solution to Einstein's equations in a lower-dimensional case—Flatland—where this third possibility applied. A time machine of my type could form, hidden inside a rotating black hole; a Flatlander could travel back in time within the black hole and then later emerge into a different universe.

If, despite the fact that you can never brag afterward, you want to build a time machine that allows you to circle the string loop once and travel back in time by about a year, you need that enormous cosmic string loop, with a mass more than half that of our galaxy. Assume your supercivilization has manipulated the loop until it forms a slightly turning, approximately rectangular geometry whose horizontal sides are about 54,000 light-years long and whose vertical sides are about 0.01 light-years tall. The rectangle will contract as the vertical sides are pulled toward each other, which causes them to lengthen. When, 27,000 years after the start, the vertical sides are half a light-year tall, horizontal segments only about 10 feet long will join them at the top and bottom. The two relatively straight vertical sides then approach and pass each other, 10 feet apart, each moving at more than 99.999999996 percent of the speed of light. Follow one of the straight string segments in at this speed and you'll only age 3 months during the 27,000-year trip.

At the point the two string segments pass each other, if possibility 2 applies, then you will be able to circle around the two cosmic string segments as they pass and go back in time about a year. But at that point you will already be inside the event horizon of the black hole that is forming as the loop completes its collapse. So you will never get back.

Now your main worry is to avoid hitting a singularity. So you may want to do more time travel to the past while inside the black hole. Say you circle the strings 11 times before giving up. What will the whole trip look like to you? When you first arrive, just as the two sides of the string loop are passing each other, you will see 11 older versions of yourself waiting there to see you off. The first one, which looks 1 year older than you, says, "Hi, I've been around the strings once!"; the second one, which looks 2 years older, says, "Hi, I've been around the strings twice!"—and so forth. After being greeted 11 times, you will go around the strings yourself, and when you return, you'll find yourself doing the exclaiming. You'll continue going around until you return for the 11th time and say, "Hi, I've been around the strings eleven times!" Then, realizing that this can't go on forever—there isn't room in the black hole for an infinite number of copies of yourself—and knowing that you did not return again, you will give up circling the strings and go on to the future, inexorably to that singularity.

This result recalls an interesting parallel. Long before my paper, British physicist Brandon Carter had investigated the geometry inside a simple, unperturbed, rotating black hole, such as might be formed by the collapse of a rotating star. There the singularity is not a point but a small ring, which in turn leads to universes beyond, according to Einstein's equations. In other words, if you traveled inside such a rotating black hole, you could jump through the ring and enter another universe. You could also avoid going through the ring and still emerge, in yet a different universe. This would be like taking an elevator that only goes up. You get on and the elevator doors close behind you—no returning to your friends on the ground-floor universe. You can observe the entire future history of the ground-floor universe as you ascend to the second-floor universe. You can leave the elevator and visit the second-floor universe—it's different from the one you started on. If you return to the elevator (go back inside the black hole), you can visit the third-floor universe, and so on. In principle, you can visit an infinite number of different universes.

But Carter discovered more. (Investigating these solutions is like making a quilt: you keep sewing on pieces, obeying the pattern, and you see what they produce.) Inside the rotating black hole, near the ring singularity, the spacetime is so twisted that you could jump through the ring and fly around parallel to its circumference to travel back in time. This creates a region of time travel trapped inside the black hole—yet one more way in which Einstein's equations allow time travel.

Any photons falling into the black hole from our universe would become very blueshifted and energetic, however. You would encounter these photons, which could kill you, as you headed in to jump through the ring singularity. Theoretically, photons entering the black hole in the infinite far future could become infinitely blueshifted and create their own singularity, blocking your way to the time-travel region. Yet work by physicists Amos Ori of Caltech and Lior Burko of the Technion-Israel Institute of Technology shows that passing through this singularity to the region of time travel just *may* be possible after all, because the singularity formed by the incoming photons would be weak. First, we expect all infinities in the curvature to be "smeared out" by quantum effects, so that the curvature would just rise to a very high but finite value. (We refer to this as "near-infinite.") Second, the buildup of curvature occurs so quickly that the associated tidal forces might not tear you apart; they simply wouldn't have time to move your head and feet much while you passed through. This would be akin to going over a speed bump very fast in your car: you would get a big jolt, but you could survive. To know the exact details of the process, we need a theory of quantum gravity, which we have not yet discovered. As Kip Thorne says in *Black Holes and Time Warps*, an astronaut "will survive, almost unscathed, right up to the edge of the probabilistic quantum gravity singularity. Only at the singularity's edge, just as he comes face-to-face with the laws of quantum gravity, will the astronaut be killed—and we cannot even be absolutely sure he gets killed then, since we do not really understand at all well the laws of quantum gravity and their consequences."

Still another time-travel possibility exists. In 1976 physicist Frank Tipler, now at Tulane University, found that if you have an infinitely tall cylinder rotating at nearly the speed of light on its surface, you could go back in time by flying around the cylinder. This solution is reminiscent of mine, with the two infinite cosmic strings passing each other.

Tipler, and later Hawking, then proved some theorems suggesting that in certain cases you would create singularities by trying to form a time machine within a finite region where none existed before. (Tipler figured that, although the universe might be infinite, human beings could control only a finite region.) Tipler knew that if you create a time machine where none existed before, you must cross a Cauchy horizon to enter the region of time travel. Tipler then examined what the structure of this Cauchy horizon would be like if the mass-energy density was never negative. If the Cauchy horizon extends infinitely, no problem exists. But if the horizon is finite, Tipler showed it must end in the past somewhere in a singularity. Thus, as you crossed the Cauchy horizon, you could look out into the past along the Cauchy horizon and see a singularity. It was generally thought that such a singularity could, in principle, spew out all kinds of elementary particles that could kill you, but the argument has a loophole: there is a singularity we already see when looking out into the past that doesn't kill us—

the big bang singularity at the beginning of the universe. Looking at a singularity isn't necessarily fatal.

Intriguing.

But seeing a singularity is only one problem. If there is only normal positive-density material present (all the stuff we are used to seeing such as protons, neutrons, electrons, and electromagnetic radiation), then the Cauchy horizon would appear to be unstable. That is, any perturbing wave circulating along the Cauchy horizon would keep building in intensity until conditions there become quite unpredictable. A supercivilization might try to actively manage the instability. (For example, a pencil balancing on its point is unstable, but if you are fast and smart, you can support the point with your hand and, by moving back and forth, keep the pencil upright. Some designs for modern fighter planes intentionally make the plane unstable in flight to increase its maneuverability, relying on computer control to actively manage the instability.) For the black hole case, this is difficult to achieve in practice. If an instability occurred, then a singularity might indeed block a time traveler's way across the Cauchy horizon—not just a singularity she would see (in the distance) but one she would actually hit. As mentioned before, to know whether our astronaut would be able to crash through the singularity speed bump into a region of time travel, we need a theory of quantum gravity. In any case, it certainly looks like a dangerous trip.

But there are further possibilities for making time machines, ones that get around some of these difficulties.

WORMHOLES

In 1988 Kip Thorne and his Caltech colleagues Mike Morris and Ulvi Yurtsever showed how time travel to the past might be accomplished by taking advantage of wormholes. As you learned

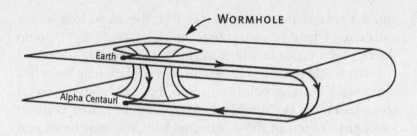

A wormhole creates a shortcut from Earth to Alpha Centauri.

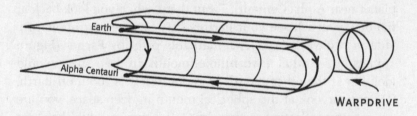

A warpdrive creates a U-shaped distortion in spacetime, also creating a shortcut from Earth to Alpha Centauri.

Figure 15. Wormhole Geometry and Warpdrive Geometry

in Chapter 1, wormholes are tunnels connecting two regions of spacetime distant from each other. Think of the wormhole in an apple; the worm can get from one side of the apple to another more quickly by going straight through its wormhole than by crawling along the curved surface of the apple. We might find a wormhole with one mouth near Earth and the other near Alpha Centauri (top, Figure 15). One could then get to a planet near

Alpha Centauri in two ways: (1) take the usual long route, extending 4 light-years through ordinary space, or (2) jump through the wormhole, which might be a trip of only 10 feet.

What would this wormhole look like? A black hole looks like a big black bowling ball (jump into that ball, and you won't come back), but the wormhole (providing its tunnel is short) looks just like one of those mirrored balls you sometimes find in a garden, reflecting the entire landscape around it. It will not be your "earthly garden," however, that you would see revealed in the wormhole ball, but one near Alpha Centauri. Jump into that ball, and like Alice in Wonderland, you would find yourself tumbling out somewhere quite different—in a garden on a planet near Alpha Centauri. From there, when you look back at the ball, you will see your earthly garden. The wormhole provides a two-way portal. A remarkable print by Escher (Figure 16) lets us see what a wormhole "mouth" in deep space would look like if the other mouth were located in a room on Earth. (When you look at the spherical mouth in deep space, you are not seeing a reflection; rather, you are seeing through the short wormhole tunnel into the room on Earth, getting a distorted view of it.) Escher drew this picture in 1921, long before Thorne and his colleagues had even proposed wormholes.

A light beam takes about 4 years to reach Alpha Centauri from Earth when traveling through ordinary space, but you could beat a light beam to Alpha Centauri by taking a shortcut through the wormhole. As in the case of cosmic strings, any time you can beat a light beam by taking a shortcut, the possibility of time travel to the past opens up.

If you found an Earth–Alpha Centauri wormhole, you could dive from Earth through it in the year 3000 and emerge at Alpha Centauri. But when? You might emerge not in the year 3000 but perhaps in the year 2990 instead. If you emerged in the year 2990 at Alpha Centauri, you could travel back to Earth at 99.5 percent of the speed of light and arrive back at Earth

Figure 16. *The Sphere* (1921), by M. C. Escher.
How a wormhole mouth might appear.

approximately 4 years later—in 2994. Thus, you would arrive back at Earth 6 years before you left. You could wait on Earth for those 6 years, so you could shake hands with yourself when you took off in the year 3000. You would have accomplished time travel to an event in your own past.

But suppose instead that the two mouths were synchronized. (Since Alpha Centauri and Earth are not moving at high speed relative to each other, observers in the two places could synchronize their watches and agree on the time.) When you jumped down the wormhole mouth on January 1, 3000, you would emerge at Alpha Centauri also on January 1, 3000. No time travel there. Thorne and his colleagues showed that the two mouths could be desynchronized by dragging around the wormhole mouth next to Earth at a speed close to that of light. This could be done by bringing a massive spaceship close to the mouth and simply letting the mouth fall by gravity toward the spaceship. When the rockets were fired to accelerate the ship, the wormhole mouth would follow like a faithful puppy. In this way, you could force the mouth to move at speeds up to 99.5 percent of the velocity of light. Starting on January 1, 3000, if you took the wormhole mouth on a trip to a point 2.5 light-years away, making a round trip at a speed of 99.5 percent that of light, observers on Earth would see that 5-light-year round trip taking just over 5 years—with the mouth arriving back on January 10, 3005.

Consider an astronaut sitting in the middle of the wormhole tunnel with a clock. Observers on Earth would see her clock ticking very slowly—10 times more slowly than theirs, because they see her traveling back and forth with the moving mouth at 99.5 percent of the speed of light. (Recall that special relativity tells us that such moving clocks tick more slowly. A clock moving at 99.5 percent of the speed of light on such a round trip ticks at a rate one tenth as fast as one on Earth, because Einstein's factor $\sqrt{[1 - (v/c)^2]}$ is one tenth in this case.) When the wormhole returns to Earth, the astronaut will have aged only half a year since the start—that's 5 years divided by 10. Meanwhile, the wormhole mouth near Alpha Centauri will not have moved, since nothing has been pulling on it. Furthermore, through this whole trip, the length of the wormhole tunnel would not change—it's still just 10 feet long. Since the mass and energy in the wormhole tunnel do not change, Einstein's equations tell us that its geometry will not change either. It will maintain the same length—only the locations it connects will change. Wait until the wormhole mouth next to Earth returns —it is now January 10, 3005, on Earth; you can jump in the wormhole, travel 5 feet, and meet the astronaut sitting in the middle. She will have aged only 6 months during the trip, so her clock says July 1, 3000. Then when you travel another 5 feet, you will emerge near Alpha Centauri and find that it is July 1, 3000, there as well. Why? Because the astronaut as seen from the Alpha Centauri end is not moving, and her clock, which has logged 6 months since the start, remains synchronized with clocks on Alpha Centauri. Emerging on July 1, 3000, near Alpha Centauri, you can now get in a regular rocket ship and return to Earth by the regular way, through ordinary space. Traveling at 99.5 percent of the speed of light, you can get there in just over 4 years, arriving back on Earth on July 8, 3004. You would arrive back almost 6 months before you started your trip. Just wait patiently on Earth until January 1, 3005, and you can shake hands with yourself as you leave—you can visit an event in your own past.

In this case, as in the moving cosmic string case, there is an epoch before which time travel cannot exist. If you lived on Earth in the year 3005, you could use the time machine to visit Earth in 3004, but not the year 2001, because that was before the time machine was created. Someone on Earth in the year 2001 would see no time travelers, but an observer on Earth in the year 3004 might well encounter them. After the wormhole mouths have been desynchronized sufficiently, time travel would be possible. But still later, in the year 3500 perhaps, if we were to move the wormhole mouth on Alpha Centauri, we could resynchronize the two mouths, causing the epoch of time travel to end. We could in this way destroy the time machine after having built it. You can use the time machine only while it is in existence.

Exotic material would be needed to prop a wormhole open so that a traveler could pass through it. Light beams converging on the wormhole mouth near Earth pass through the wormhole and spread apart as they exit the wormhole near Alpha Centauri. This is the hallmark of the repulsive effects of negative-energy-density stuff. You would have to add energy to this to return to zero. Surprisingly, there are quantum effects that actually produce a negative energy density. Thus, Thorne and his associates hope that a supercivilization in the future might be able to use such quantum effects to keep a wormhole open. Another problem to be solved is how to attach the wormhole mouths where you want them. Perhaps there are already microscopic wormholes 10^{-33} centimeters across, connecting many places and times in spacetime. Some supercivilization might be able to enlarge one of these microscopic wormholes so a spaceship could pass through it.

Because wormholes are propped open by negative-energy-density stuff, they are stable, avoid the singularities implied by Tipler's theorem, and can create a time machine without the danger of forming a black hole. However, they are still subject to quantum effects that may interfere with their operation, a point I consider again in Chapter 4.

WARPDRIVE

The wormhole has a sister possibility in travel: warpdrive. In *Star Trek*, the crew of the *Enterprise* used warpdrive to alter space so they could travel among the stars at speeds faster than that of light. Welsh physicist Miguel Alcubierre took this idea seriously and showed how a warpdrive could work, using the principles of general relativity. In this case, you could take a 4-light-year path from Earth to Alpha Centauri and warp that space so that the distance through the resulting tube would be just 10 feet.

Picture it this way. Imagine yourself as an ant living on top of a dining room table. You want to visit another ant living on the bottom of the tabletop. To visit your friend, you could crawl 2 feet to the nearest edge of the table, walk a half inch down the edge, and then scurry 2 feet back to the center of the table's underside. The total distance traveled would be 4 feet and 1 half inch. Or you could drill a hole through the table and crawl down half an inch through the hole and visit your friend directly—a veritable wormhole. A third way to reach the table's underside would be to use a jigsaw to cut a 2-foot-long slit in the table between the table edge and the table's center. This way you could also crawl down half an inch, this time through the slit, to visit your friend. Your journey would still be only half an inch. To other ants crawling around the tabletop, the table looks the same, provided they never venture near the slit. If the table were made of soft rubber, it could be deformed, or warped, to produce a slit without cutting. Just press very hard on the edge of the rubber tabletop until you have pushed 2 feet in toward the center. Without changing the table's topology (you have not drilled any holes through it), you have changed the shape of the tabletop by warping it. That's how warpdrive gets its name.

If you examine Figure 15, you will see the similarity between wormhole geometry (top) and warpdrive geometry (bottom). To produce the geometry necessary for a warpdrive shortcut from Earth to Alpha Centauri, Alcubierre found that both some ordinary positive-energy-density stuff and some exotic nega-tive-energy-density stuff are required. With his solution, a tubular path warped in spacetime would get you to Alpha Centauri quickly, and another warp could allow you to return to Earth quickly as well. You could go to Alpha Centauri and be back in time for lunch on Earth the same day. Alcubierre recognized that since his warpdrive solution allowed you to beat a light beam, future refinements might lead to a solution allowing time travel to the past, but he did not say how to do it. Soon after, using an argument similar to the one I used for cosmic strings, Allen Everett of Tufts University showed that he could construct two moving warpdrive shortcuts that would allow time travel to the past.

Thus it appears that Gene Roddenberry, the creator of *Star Trek*, was indeed right to include all those time-travel episodes! Unfortunately, however, Russian physicist Sergei Krasnikov showed that the *Enterprise* could not actually create its own warpdrive path to wherever it wanted to go, as it does on the show. The path would need to be laid out in advance by ships going slower than the speed of light. The *Enterprise* would be more like a train traveling along prelaid tracks than an all-terrain vehicle venturing out alone. A future supercivilization might want to lay down warpdrive paths among stars for starships to traverse, just as it might establish wormhole links among stars. A network of warpdrive paths might even be easier to create than one made up of wormholes because warpdrives would require only an alteration of existing space rather than the establishment of new holes connecting distant regions.

DIFFICULTIES WITH TIME TRAVEL TO THE PAST

Since all the proposed methods of traveling to the past have their own difficulties, let's consider another idea for communicating with the past: tachyons. These are hypothetical particles that travel faster than the speed of light.

What!? We've already agreed that nothing can travel faster than light. True, normal particles like those you and I are made of (protons, neutrons, and electrons) must move slower than the speed of light; otherwise, Einstein's postulate that all observers should be able to think of themselves as at rest would be violated. And photons always travel at the speed of light through empty space.

But let's imagine, as physicists S. Tanaka, O. M. P. Bilaniuk, V. K. Deshpande, and E. C. G. Sudarshan did in the early 1960s, a particle that always travels faster than light. American physicist Gerald Feinberg called such a particle a tachyon, after the Greek word *tachys*, meaning "swift." Since tachyons can beat light beams back and forth, with the help of an astronaut friend, one could use tachyons to send a signal into one's past. This was the basic idea Gregory Benford used in his 1980 science-fiction novel *Timescape*. Would it work out in reality?

Tachyons can be made compatible with special relativity, but the equations of general relativity create dilemmas. A tachyon would have to be accompanied by gravitational waves, just as an airplane exceeding the speed of sound creates a sonic boom. In 1974, using a 1972 result found by F. C. Jones plus my own solution to Einstein's field equations for a tachyon in a different context, I found that a tachyon should emit a cone of gravitational radiation that would trail behind it. The emission would cause the tachyon to lose energy and, because of the hypothetical particle's peculiar nature, cause it to accelerate to still higher speeds. In keeping with Jones's insight, the particle's world line through spacetime is bent like a wide arch. We would see the two sloping sides of the arch as a tachyon and an anti-tachyon approaching each other at just over the speed of light, going faster as they got very close, finally reaching infinite speed as they hit and annihilated each other at the top of the arch. After that, no more tachyon. Because tachyon world lines would bend in this way, tachyons would spend most of the time moving at just barely over light speed. Therefore, tachyons could not be used to send energy or information faster than light over macroscopic distances.

A final proposal for traveling back in time makes use of antiparticles.

Late one night, John Wheeler, at Princeton, called Richard Feynman and exclaimed, in effect, "I know why all electrons have the same mass—they are all the same electron!" Wheeler's idea was that a positron (the "antiparticle"—particle of identical mass but opposite charge—paired with the electron) could be thought of as an electron traveling backward in time. To understand this idea, imagine tracing a large capital letter N in a spacetime diagram in which the future is up, the past is down, and space is horizontal. Starting at the bottom left, move your finger upward on the first stroke of the N to trace the path of an electron moving toward the future. Then move your finger down diagonally to trace the path of a positron, which could be interpreted as an electron going backward in time. Finally, trace the last line upward; that's an electron again. This N-shaped world line in spacetime plays out to us as the following movie: scanning a horizontal ruler up slowly from the bottom of the N to the top, we see one electron sitting still on the left and the creation of an electron-positron pair on the right; the positron travels from right to left, finally meeting up with the electron on the left, at which point they annihilate each other. We might, of course, simply interpret the N as three particles—two electrons and a positron—all moving forward in time. Wheeler thought that all the electrons in the universe might be part of one long world line that zigzagged forward and backward in time many times. Each "zig" would look like another electron, and each "zag" would be a positron. The corners between the zigs and zags would look like either the creation or the annihilation of an electron-positron pair.

For this idea to work, the number of positrons and electrons in the universe at any time would have to be nearly equal. Unfortunately, many more electrons seem to exist in the universe in the present epoch than do positrons. Nevertheless, the idea that positrons can be considered electrons traveling backward in time seems valid and was ultimately used by Feynman in his diagrams for quantum electrodynamics, for which he received the Nobel Prize.

For you to use this method to travel back in time—to create for yourself an N-shaped world line—would require that incredibly improbable events occur. First, near your location, out of the energy of several thousand H-bombs, a highly organized "you–anti-you" pair would have to be created. These two doppelgängers—the "anti-you" representing your backward zag in time, and the other "you" representing your resumption of forward motion in time—would each have to reproduce you exactly, down to the atomic level. Then the anti-you would have to come over to meet you. Arranging for each particle in the anti-you to annihilate each corresponding particle in you, so that the energy produced would not disperse the structures in your body before the annihilation could be completed, would seem impossibly difficult. So if you see an antimatter version of yourself rushing toward you, think twice before embracing.

Time travel to the past would appear difficult at best, so don't call your travel agent just yet. Extreme conditions would be required to even attempt such a project. Time machines for visiting the past are not something you will be building in your garage, à la the first Apple computer. They are, as Kip Thorne has noted, at best, projects for supercivilizations of the future.

But physicists spend their energy exploring the possibilities of time travel in principle for a very good reason: as I've commented earlier, we are interested in testing the boundaries of the laws of physics under extreme conditions. As physicists often remark (especially when they lack money for building a new particle accelerator), in its early moments the universe itself was a particle accelerator. If we want to search out places where extreme conditions prevail, we can look to the interiors of black holes—or to the beginning of the universe. Stephen Hawking, early in his scientific career, showed how some theorems about the singularities occurring in the centers of black holes could be applied to the early universe. Through studies of modern inflationary cosmologies, we are now seeing that the early universe should have had event horizons just as black holes do, separating us from distant regions that are forever beyond our view. Improving our understanding of the physical parameters of black holes could thus help us appreciate what happened in the early universe.

A similar logic applies to time machines. If we wish to test whether the laws of physics allow time travel to the past, we might further explore extreme situations. One place to look for naturally occurring time machines may be in the interiors of black holes. The curvature of spacetime was also extreme at the beginning of the universe—did a time machine exist there as well? If so, it might explain how the universe got started in the first place.

4 TIME TRAVEL AND THE BEGINNING OF THE UNIVERSE

> Whence this creation has arisen—perhaps it formed
> itself or perhaps it did not—the one who looks down
> on it, in the highest heaven, only he knows—or per-
> haps he does not know.   —THE RIG-VEDA
>   (TRANSLATED BY WENDY DONIGER O'FLAHERTY)

A LETTER FROM LI-XIN LI

One day at my Princeton office I opened a letter. It was from Li-
Xin Li, a student from China. He was interested in coming to
Princeton to study for a Ph.D. in astrophysics and work with
me on time travel. He included a paper he had written on the

subject. It was not unusual for prospective students to send in letters, or even to enclose papers. I typically just forward these to the graduate school for consideration by our department at its admissions meeting. But this case was different because Li-Xin Li's paper was already well known to me. I had read it and particularly liked it when it had been published in the *Physical Review*. It addressed a problem raised by Stephen Hawking—that quantum effects might always conspire to prevent time travel. The particular example concerned time travel using a wormhole; waves circulating between the two wormhole mouths might build up an infinite density—a singularity —in the quantum state, potentially shutting down the time machine before it started. Li-Xin Li proposed the ingenious solution of putting a reflecting sphere between the two wormhole mouths to reflect the waves and stop the infinite buildup of energy. I had never received such an important paper from a prospective student. It proved that he was one of the few dozen people in the world able to do these complex quantum calculations—and furthermore, that he had original ideas. Even more to the point, he was interested in time travel.

I was reminded of stories I had heard as a postdoc at Cambridge University, in 1975, about how Professor G. H. Hardy received a letter from a young man named S. Ramanujan, from India. The correspondence included some remarkable theorems he had proven. These tales were related to me by Hardy's friend, the famous mathematician J. E. Littlewood, who, then in his nineties, was our senior fellow at Trinity College—Isaac Newton's old college. It's a pretty heady place. You can walk down the corridor where young Newton clapped his hands in rhythm with his handclap echoes from the corridor's far end, to measure the speed of sound. Dinner every night is in the Great Hall, after which the fellows retire upstairs to drink port, smoke cigars, and pass the snuffbox—an experience not unlike being transported back in time by a time machine. The senior fellows entertain their junior colleagues with stories about Trinity people, who over the years have also included Alfred Lord Tennyson, Lord Byron, and James Clerk Maxwell. As an undergraduate, Byron used to keep a pet bear tethered to the fountain in the Trinity courtyard, and Newton stayed away from Trinity during the plague years, when he developed calculus and first thought of gravity extending to the orbit of the Moon.

Hardy had shown Ramanujan's letter to Littlewood, saying such theorems must have come from a mathematician of the highest order. So Ramanujan was invited to come to Trinity College. Hardy and Ramanujan, working together, then produced a most remarkable theorem in number theory: a formula for accurately estimating the number of different ways a given sum could be achieved. (Once when Ramanujan became ill, Hardy went to visit him. To cheer him up, Hardy said, "I have just arrived in Taxicab No. 1,729. What a dull number!" "No," Ramanujan replied, "it is a very interesting number: it is the smallest number that is the sum of two cubes two ways." Indeed, it's true: $1,729 = 1^3 + 12^3 = 10^3 + 9^3$. Amazing.)

When I looked at Li-Xin Li's letter, I wondered if he might be a similarly remarkable person. In my many years as chair of the judges for the Westinghouse—and Intel—Science Talent Search, the nation's oldest and most prestigious science competition for high school students, I learned that the best predictor of future success in research is having done good research in the past. It's better than SAT scores, grades, or letters of recommendation. I thought Li-Xin Li was a great prospect. I recommended him in the highest terms to my colleagues, and—to make a long story short—we admitted him to Princeton's Department of Astrophysical Sciences with a fellowship.

I did have a good idea for him to work on: how time travel might be applied to explain the origin of the universe. But an important problem needed to be addressed. Could one find a quantum state for the early universe involving time travel that would work?

Li-Xin Li arrived several months early, and although he was not yet formally enrolled as a student, there was no reason for us not to set to work. And, if you're researching something really important, keeping it to yourself until you are finished is usually a good idea. Li-Xin Li and I met once a week for lunch, and we didn't tell anybody what we were working on. Those were memorable get-togethers. We tried out several local restaurants before settling on the Orchid Pavilion. At one lunch, early on, while we were hard at work on our theory of the origin of the universe, we received a fortune cookie that said, "Trust your intuition. The universe is guiding your life." We took this as some encouragement.

VACUUMS AND CHRONOLOGY PROTECTION

To tell the rest of this story, I need to tell you about different kinds of vacuums because they play a key role in Li-Xin Li's and my work. Not Hoovers, mind you, but the kind of vacuum that is left after you have emptied the room you're in of all the people, all the furniture, all the air. Sweep out all the elementary particles, including photons. You are then left with empty space —a vacuum. A *normal vacuum* is expected to have zero density and zero pressure. But quantum mechanics tells us that empty space may not always be a vacuum with a zero energy density. In 1948, Dutch physicist Hendrik Casimir showed that if you place two electrically conducting silver plates very close together, the empty space between them has a vacuum that has a negative energy density—that is, the amount of energy per cubic centimeter is actually less than zero. You would have to *add* energy to this to get back to zero. The *Casimir vacuum* is

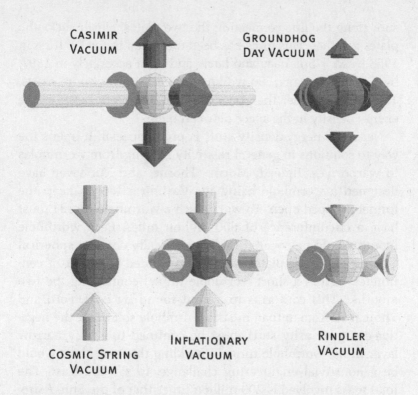

Figure 17. Different Kinds of Vacuums

illustrated in Figure 17, along with some other vacuums. In these vacuums, energy density is represented by a sphere. A lightly shaded sphere represents negative energy density; a darker sphere denotes positive energy density. The pressure in different directions is indicated by arrows. Darker arrows pointing outward show positive pressure, like the pressure in a car's tires. Lighter arrows pointing inward stand for negative pressure, or suction. The Casimir vacuum has a positive pressure in the two directions parallel to the plates but a large negative pressure along the line connecting the two plates, which sucks the plates together. This force has been measured in the lab (first in 1958 by M. J. Sparnaay and lately and most accurately in 1997, by S. K. Lamoreaux). So we know the Casimir vacuum exists. The closer together the plates are held, the more negative the energy density in the space between them.

Negative-energy-density stuff is pretty special. It opens the way to solutions in general relativity ranging from wormholes to warpdrives. Indeed, Morris, Thorne, and Yurtsever have designed a wormhole using the Casimir effect to keep the tunnel propped open. To work, such a wormhole tunnel must have a circumference of 600 million miles. Each wormhole mouth would be covered by an electrically charged spherical Casimir plate. The plates would be separated by only 10^{-10} centimeters across a short wormhole tunnel connecting the two mouths. (This conforms to a limit found by L. H. Ford and Thomas A. Roman that, in such wormhole solutions, the negative-energy-density stuff must be confined to a very narrow layer in the wormhole tunnel.) Building this wormhole would be a nontrivial engineering challenge, to say the least. The total mass involved is 200 million times that of the Sun. Astronauts wishing to pass through the wormhole would have to avoid being fried by blueshifted radiation falling onto the plates and would have to open trapdoors in each plate in turn to get through. Not easy—but the Casimir vacuum at least creates the possibility.

Vacuums are important for cosmic strings too. Inside a cosmic string there should be a vacuum state with a positive energy density and a negative pressure along the length of the string (see Figure 17), which creates a tension along the string, making it rather like a rubber band. That is what's inside a cosmic string—an unusual high-energy vacuum state.

Vacuum states play yet another crucial role in time-travel

research, one that turns out to be important in studies of the early universe as well. Stephen Hawking felt that the vacuum state might always blow up as one tried to enter a time machine, altering the geometry of spacetime, creating a singularity, and spoiling one's chances of making a trip to the past. Hawking's intuitions on this point were informed by what would happen in *Misner space*, a spacetime where no time machine originally exists but where an epoch of time travel eventually develops. The region of time travel is separated from the region without time travel by a Cauchy horizon, just as in my cosmic string case. Think of Misner space as an infinite room bounded by a front wall and a back wall. You live between the two walls. There is a door in the front wall and a door in the back wall. Go out the front door. You then find yourself immediately reentering the same room via the back door. Aha: Misner space is actually rolled up like a cylinder—its front and back walls are "taped together."

This space may make you feel a little claustrophobic. But then things get worse. You notice the two walls are approaching each other. In fact, they are moving at constant velocity and will hit each other in the future—say, in an hour. This is like being trapped in the huge garbage compacter in the original *Star Wars* movie—the walls are moving toward each other, and you are trapped between them. Real escape is possible in Misner space, however. Go out the front door; as you know, you will reenter the same room from the back door. Now go out the front door again—and keep repeating these actions. Because the walls are moving toward each other, every time you pass through the room, you pick up additional velocity with respect to the walls. You will keep circulating through the room, again and again, faster and faster. Pretty soon, the front wall will be approaching you at nearly the speed of light. Because the whole room is moving faster and faster with respect to you now, according to special relativity it will look narrower and narrower to you each time you pass through it. Besides, the walls are actually getting closer together with time. Because of these effects, you can actually pass through the room an infinite number of times in a finite time as measured by your watch. Where do you go then? You then cross a Cauchy horizon into a region of time travel. You are not in the room anymore. You're not in Kansas anymore. You've entered a peculiar spacetime Oz. The new region resembles a piece of paper—the past is at the bottom of the page and the future is at the top, and you roll it up and tape the top to the bottom (as in Figure 9). You can keep visiting the same events again and again. Misner space is definitely strange—but calculating what happens in it is relatively easy. It is often taken as an archetypal example of a spacetime in which a time machine is created (as in the wormhole and cosmic string cases).

Physicists William Hiscock and Deborah Konkowski of Montana State University calculated the kind of vacuum that would apply in Misner space. They started with a quantum state corresponding to the normal vacuum and asked how it would change if it were wrapped around a room whose front and back walls were taped together. The walls, being taped together, would act like Casimir's parallel plates, so Hiscock and Konkowski found that inside the room there would be a Casimir vacuum with negative energy density. As we have discussed, you could depart this room at the front and automatically reenter at the back. As you fly through the room again and again, it would become ever more narrow. The distance between the front and back walls, the circumference of the cylinder, is getting smaller and smaller. The closer the walls move toward each other, the thinner the cylinder gets, and the more negative the vacuum's energy density would become. Finally, just as you are about to escape into the region of time travel, the negative energy density would blow up, becoming negative infinity.

This would produce infinite space curvature—a singularity—and thus might prevent you from ever getting to the region of time travel.

This finding motivated Stephen Hawking to propose his *chronology protection conjecture*—that the laws of physics always conspire to prevent time travel to the past. If the quantum vacuum always blew up, creating a singularity as you approached a region of time travel, that, coupled with the other nasty effects I've earlier mentioned, might always stop you from getting into a region where time travel could occur.

But I wanted to look at the Hiscock and Konkowski calculation again. I hoped there might be some way to remedy its difficulties—in the same way Stephen Hawking's discovery of the emission of what's now known as Hawking radiation solved some similar vacuum blowup problems near the event horizons of black holes.

I first asked Li-Xin Li to calculate the vacuum state in a simpler spacetime involving time travel, which I call the *Groundhog Day spacetime*. In the movie *Groundhog Day*, as I mentioned in Chapter 2, Bill Murray's character keeps reliving the same day, which happens to be Groundhog Day. Each night he goes to bed and sleeps until his alarm sounds at 6:00 A.M. To his dismay, he discovers that it is 6:00 A.M. on Groundhog Day again, and he is right back where he started. The Groundhog Day spacetime is created by simply taping 6:00 A.M. Tuesday and 6:00 A.M. Wednesday together to form a cylinder. (See Figure 9.) In this spacetime, when you get to 6:00 A.M. Wednesday, you simply find yourself back at 6:00 A.M. Tuesday. Your world line could then be a helix wrapping around and around the cylinder as you relive the same day over and over. If you lived for 80 years (29,220 days), your world line would wrap around the cylinder 29,220 times, and as you aged you would encounter 29,219 other copies of yourself, ranging from babies to senior citizens.

In this spacetime scenario, you could play football against

yourself—in fact, you could play each position on both teams in turn, and you could be all the spectators as well. You could go to the stadium and play quarterback for one team, go back in time to play quarterback for the other team, and so forth, ending by going to the stadium as a spectator and sitting in a different seat each time. It would seem like many days to you, but you would really be witnessing the same events over and over. That football game would always have the same outcome—because it would be one game.

Li-Xin Li found that a normal vacuum wrapped around the cylindrical Groundhog Day spacetime would have a positive energy density and a positive pressure (Figure 17). The energy density and pressure would be small and so would not greatly alter the geometry. No infinite blowup of energy density would occur. Quantum vacuum blowups never appear to interfere with time travel in situations where the time travel has always been present. Groundhog Day spacetime has time travel throughout—every event can be visited again. It has no Cauchy horizon dividing a region of time travel from a region of no time travel. Yet the vacuum Li-Xin Li found for Groundhog Day spacetime was quite like the one that Hiscock and Konkowski had found within the time-travel region of Misner space.

I then asked Li-Xin Li to calculate the normal wrapped vacuum in Misner space just as Hiscock and Konkowski had done. He got the same results they did. Was there any time travel solution that would work?

The next time we had lunch, Li-Xin Li said, "I have the answer." He noted that in a given geometry there is more than one vacuum state to choose from. Instead of starting with the normal vacuum, he started with a kind called the *Rindler vacuum.*

The Rindler vacuum is the vacuum state measured by accelerated observers. To understand it, one must first understand that an accelerating astronaut firing his rocket ship in empty space with a normal vacuum will, surprisingly, detect photons. This thermal radiation is called *Unruh radiation*. You don't see it if you're not accelerating, but an accelerating astronaut does. Where do these photons come from? In effect, their energy comes from borrowing against the normal vacuum, like someone putting real money in his pocket by taking out a loan and going into debt. The energy the astronaut "borrows" from the vacuum therefore makes him observe a vacuum with an energy density below zero—a vacuum state called the Rindler vacuum. The Rindler vacuum has negative energy density and negative pressure (see Figure 17). This just counteracts the positive energy density and positive pressure of the Unruh radiation that the astronaut observes, making the total energy density and pressure add up to zero and agreeing with the normal vacuum state seen by a nonaccelerating observer such as you. The astronaut detects photons; you do not. You and he disagree about what the vacuum state is and whether any photons are present, but you both agree on the total energy density. The normal vacuum you see is equal to the Rindler vacuum he sees plus the Unruh radiation he detects. If the accelerating observer saw no radiation, he would conclude that the energy density was actually less than zero and that he was living in a universe with a pure Rindler vacuum. The Rindler vacuum is well known (to physicists, at least) for describing the vacuum state for accelerated observers. (For additional details on this vacuum, see the Notes.)

A Rindler vacuum in the time-travel region sets up a negative energy density and a negative pressure. But then, because the spacetime is wrapped around in the time direction, positive energy density and pressure are added to this (as occurred in the Groundhog Day spacetime). With appropriate parameters, the two effects neatly cancel each other, leaving a vacuum with zero energy density and pressure, just like the normal vacuum. For this cancellation to occur, the front and back walls in the Misner space must approach each other at a velocity of 99.9993 percent of the speed of light.

This was a beautiful solution. Li and I shook hands. This wrapped Rindler vacuum had zero energy density and pressure throughout the entire Misner space—in both time-travel and non-time-travel regions—and therefore it solved Einstein's equations exactly. This was a self-consistent solution: the geometry, which included time travel, gave rise to the quantum vacuum properly, and that quantum vacuum state, through Einstein's equations, in turn produced the geometry one started with. This solution provided an important counterexample to the chronology protection conjecture, for it concerned the very example that had helped motivate the conjecture in the first place.

Li-Xin Li and I immediately knew that the solution could be adapted to produce a self-consistent state for the model of the early universe involving time travel that we were working on. The next step was to show that it would work. It did. Our paper on cosmology would take months to finish, so we decided that we should write a separate paper on Misner space and send it quickly to *Physical Review Letters*. Li-Xin Li would be first author on that paper, as he had made the crucial breakthrough. We sent our paper off on September 5, 1997. We put one cryptic sentence in the paper, noting that we had also a self-consistent solution for a kind of space that someone might have realized was a solution for the early universe involving time travel. We hoped this sentence would establish our priority without giving away the whole idea. Meanwhile, we worked furiously to complete the calculations for our paper on the creation of the universe.

In the November issue of *Classical and Quantum Gravity*, a new paper appeared by Michael J. Cassidy, Stephen Hawking's student. He had proved that a vacuum state must exist for Misner space having a zero density and zero pressure throughout. He had deduced this by reasoning from the vacuum existing around a cosmic string. He didn't know what this state was; he just knew it must exist and must be different from the one Hiscock and Konkowski had used. Furthermore, this state occurred when the front and back wall approached each other at 99.9993 percent of the speed of light. The quantum vacuum state he had shown must exist was clearly the one we had already found! And he had beaten us into print. We now felt that it would be just a matter of time until someone else found our solution, so we immediately put our Misner space paper onto the website for astrophysics preprints (xxx.lanl.gov/astro-ph) so anyone could read it. We then redoubled our efforts to get our cosmology paper out as quickly as possible, taking only Christmas Day off. We submitted it online to the *Physical Review* and posted it the next day on the Internet just as the year was running out—on December 30, 1997. Our Misner space paper was published in *Physical Review Letters* on April 6, 1998, and our cosmology paper—entitled "Can the Universe Create Itself?"—was published online in the *Physical Review* on May 29, 1998, and appeared in the print issue of July 15, 1998. In our cosmology paper, we used the idea of time travel to address one of the oldest and most perplexing problems in cosmology—the question of first cause.

THE QUESTION OF FIRST CAUSE

The dilemma of first cause has troubled philosophers and scientists alike for over two thousand years. Causes precede effects. If you designate a first cause for the universe, then a skeptic might ask, "So what caused that? What happened before that?" Aristotle proposed that the universe existed eternally in both the past and future—no need to ask those vexing questions. This type of model has been attractive to modern scientists as well. Newton envisioned a universe infinite in space (otherwise, he thought, the universe would collapse to a single mass) and eternal in time. When Einstein developed general relativity and applied it to cosmology, his first cosmological model was the Einstein static universe, which also lasted forever, having no beginning and no end.

In Einstein's static universe, the volume of space was finite because space curved back on itself. We may understand this by looking at an analogous situation: Earth's surface, which also curves back on itself. Earth's surface has a finite area but no edge. Columbus showed that if you sailed west, you would not fall off the edge of Earth, and Magellan's crew proved that you could keep sailing all the way around Earth and return to where you started.

The surface of a sphere is a two-dimensional surface—latitude and longitude suffice to locate a position. Imagine a Flatlander living on the surface of a sphere. He would not be able to see off the surface of the sphere. But the Flatlander could discover that he lived on the surface of a sphere, rather than on a flat plane, by noting that every time he went on a trip in what he thought was a straight line, he would return to where he started. If he brought out his surveying instruments, he would discover that the sum of angles in a triangle was greater than 180 degrees. In other words, his world did not obey the laws of Euclidean geometry. The Flatlander could even construct a triangle with three right angles by connecting the North Pole to a point on the equator with a north-south line, then going a quarter of the way around the equator, and finally turning north to return to the North Pole. You could never make such a triangle on a flat plane. Our Flatlander could discover that he lived not in Flatland, but rather in Sphereland (the plot in a novel of that name by D. Burger). Similarly, a one-dimensional Linelander could find himself living not in Lineland but in Circleland. He would travel always to the right, but after going around the circumference of the circle he would return to his point of origin.

A circle is a one-dimensional version of a sphere. Sometimes mathematicians call it a *one-sphere*. An ordinary spherical surface, such as the surface of a bubble, is called a *two-sphere* because it is two-dimensional. In Einstein's static universe, the geometry of space is a *three-sphere*—the three-dimensional counterpart of the two-sphere. In a three-sphere you will find that you live in a three-dimensional space that curves back on itself. If you travel straight ahead in your rocket ship and just keep going, you will eventually return to your home planet— approaching it from the back. What you think is a straight line in the forward direction is actually a circle with a finite circumference. Travel in your rocket ship straight to the right, and you will eventually return to your home planet from the left. Go straight up, and you will eventually return home from the bottom. No matter which direction you go, you will return home after traveling a distance equal to the circumference of the three-sphere. This three-sphere universe could be very large, having a circumference of perhaps 10 billion light-years. That way, traveling at a speed slower than the speed of light, it would take you more than 10 billion years to travel around the universe and return home.

Einstein's static universe is illustrated in the spacetime diagram in Figure 18. It shows only one of three dimensions of space in addition to the dimension of time. Einstein's static universe looks like the surface of a cylinder. Time goes upward toward the future in the diagram, and one dimension of space goes around the circumference. If you want to know what the universe looks like at a given time, then cut a horizontal cross section through this surface—you will get a circle. That circle

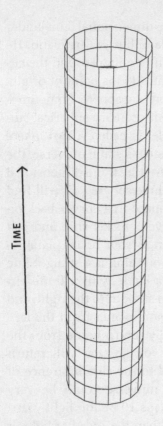

TIME →

Figure 18.
Einstein's Static Universe

represents a great circle on the three-sphere. If you want a movie of how this universe evolves with time, imagine a horizontal plane cutting the cylinder and imagine that plane sweeping upward with time. You will see a circle that stays constant in size over time: the three-sphere universe is static, neither expanding nor contracting —its circumference is always the same. The cylinder extends infinitely toward the past and future. As Aristotle would have wanted it, the universe was never created and would never be destroyed. It just exists eternally.

But unlike Newton's universe, in Einstein's static universe space is not infinite—space closes back on itself and is finite. Only the surface of the cylinder is real. Both the inside and outside of the cylinder do not exist. The world lines of galaxies (this universe has a finite number of galaxies) are the straight vertical lines going from the bottom of the cylinder to the top. These are their world histories through time. The galaxies remain at constant distances from one another. If you measured the distance between two galaxies at a given time and checked again later, the distance would be the same. Interestingly, no individual galaxy occupies a special position. Just as all points on a sphere are equivalent, all points in a three-sphere are equivalent. There is no special galaxy that can call itself the center.

To make this model, Einstein had to alter his equations of general relativity. Stars attracted one another in his theory, and so even if one started with a universe at rest, it would immediately start to collapse. A static model like that favored by Isaac Newton was not possible. Newton argued that in an infinite universe, even though stars might all attract one another, each star would have an equal number of stars pulling it in different directions, so it would just stay put. (Neither Newton nor Einstein knew about galaxies, but the argument was the same whether one talked of stars or of galaxies.) It was a trick (now questioned) that appeared possible in Newton's theory because he had a notion of absolute space and absolute time. But this option was not open to Einstein. His equations show emphatically that an infinite static universe containing stars (and galaxies) lasting forever, as proposed by Newton, was not possible. In fact, no static model was possible.

So Einstein added a new term to his equations, calling it the *cosmological constant*. Einstein expressed this term as an extra curvature that empty spacetime would always possess. In modern terminology, this is equivalent to proposing a quantum vacuum state with a positive energy density and a negative pressure. Today we would call this an *inflationary vacuum state* (see Figure 17). Einstein was smart enough to figure out that, in cosmology, a vacuum state should look the same to different observers flying through it at all different speeds below the velocity of light so that the vacuum state established no unique "state of rest." That implied that if empty space had a positive energy density, it must have a negative pressure of equal magnitude as well. This negative pressure is a sort of universal suction. If you put some of this inflationary vacuum state into your car tires, they would collapse—pulled inward by the negative

pressure. But if all of space were permeated by a constant negative pressure, there wouldn't be any differences in pressure to
push things around—so you wouldn't notice it. Similarly, in the
room where you are sitting there is an air pressure of 15 pounds
per square inch, but since it is uniform, you don't notice it.

Such a uniform pressure has one effect, however. Einstein's
equations tell us that pressure produces gravitational effects
(something not anticipated by Newton). A positive pressure,
such as might occur in a star, produces a gravitational attraction, so a negative pressure must produce a gravitational repulsion. Since there are three dimensions of space, the negative
pressure of the inflationary vacuum operates in three directions, making the repulsive gravitational effect of the negative
pressure three times larger than the attraction produced by the
energy density of the vacuum. Thus, the inflationary vacuum
state produces an overall gravitational repulsion.

This repulsion could counteract the gravitational attraction
of the stars and galaxies to allow a static universe. Moreover,
according to Einstein's calculation, if the average density of the
universe's ordinary matter (stars and galaxies) was low, then
the energy density in the vacuum would be low also and the
circumference of the three-sphere universe would be large. For
example, if the mean density of the stars and galaxies smeared
out over all space was equal to about 280 hydrogen atoms per
cubic meter, then the circumference of Einstein's static universe would be about 10 billion light-years. This is large enough
to make the effects of the curvature unnoticeable on small
scales—just as a little piece of Earth looks approximately flat.
In such a case we might initially believe that our universe
obeyed the laws of Euclidean geometry, when in fact it was a
closed, curved universe, just as some of our ancestors originally
imagined Earth to be flat, but later discovered it was a sphere.

Einstein's model had problems, however. It was unstable,

like a pencil balancing on its point. Such a balancing act couldn't be kept up forever. As stars burned, producing radiation, the total pressure in the universe would increase, pushing the model out of balance and precipitating a collapse.

And the problems didn't end there. In 1929, Edwin Hubble showed that the universe was expanding. When Einstein heard of Hubble's discovery, he pronounced his introduction of the cosmological constant "the biggest blunder" of his life. Why? Because if he had stuck to the original version of his theory, without the cosmological constant, his theory would have predicted, prior to Hubble's observations, that the universe must be either expanding or contracting. In that case, Hubble's discovery would have been the crowning vindication of Einstein's theory of gravity. This would have been experimental verification on a cosmic scale, far more impressive than even Einstein's successful forecast of light bending around the Sun. As it was, such a cosmological prediction actually did precede Hubble's discovery, for a young Russian mathematician and meteorologist, Alexander Friedmann, had published the correct cosmology models in 1922 and 1924, based on Einstein's original theory of gravity—with no cosmological constant. But few people knew of Friedmann's solutions prior to Hubble's discovery. However, if Einstein himself had discovered and announced those solutions in, say, 1917, everyone would have listened, and Einstein would have been carried through the town shoulder-high when Hubble's discovery was announced. Alas for Einstein, history went another way, and so we now also turn to Hubble and Friedmann.

The Big Bang

Edwin Hubble, working with the 100-inch-diameter telescope on Mount Wilson in California, made not one, but two, landmark discoveries. First, he found that our galaxy, a spinning pinwheel of some 400 billion stars, was not alone in the universe. He proved that many spiral-shaped nebulae, previously thought by some to be glowing clouds of gas within our own galaxy, were actually other galaxies like ours. He proved his case by identifying faint variable stars in the Andromeda nebula that were just like similar stars seen in our own galaxy; but these were very faint, proving the nebula was very far away. Our galaxy, the Milky Way, is about 100,000 light-years across. Andromeda, our galaxy's slightly larger sibling, is 2 million light-years away. The Milky Way, together with Andromeda and a couple dozen other smaller galaxies, form our Local Group of galaxies. Hubble found the galaxies beyond were sprinkled throughout space in all directions as far as his telescopic eye could see. He classified them by type—spirals, ellipticals, and irregulars—like some lucky biologist cataloging creatures for the first time. Just as Leeuwenhoek discovered the microscopic world, Hubble discovered the macroscopic universe.

Then even more news jarred Einstein's concept of the universe. Vesto M. Slipher of the Lowell Observatory in Flagstaff, Arizona, measured the velocities of more than 40 galaxies, finding that most were moving away from us. This is how such movement can be measured. A prism can be used to spread the light from a galaxy into a spectrum showing the different colors. Lines in the spectrum appear corresponding to light emission or absorption by specific chemical elements at specific wavelengths. If spectral features of known chemical elements were shifted slightly toward the red (long-wavelength) end of the spectrum (a Doppler shift), then the galaxy was moving away from us. Waves coming from a galaxy moving away will be stretched or lengthened (shifted to the red end of the spectrum) because of the steadily lengthening distance to the galaxy. A shift toward the blue, however, meant the galaxy was approaching us. While the Andromeda galaxy showed a blueshift, falling back toward us in a long and lazy orbit, Slipher found that redshifted galaxies far outnumbered blueshifted galaxies. Hubble investigated this further and found that the more distant galaxies were receding from us more rapidly. By 1931 he and his associate Milton Humason had found a galaxy receding from us at the astonishing speed of nearly 20,000 kilometers per second. A galaxy's velocity while moving away from us was approximately proportional to its distance from us, a relationship Hubble first noted in 1929, and in 1931 cemented with much more dramatic data extending to greater distances. The farther away a galaxy was, the smaller it would appear in the sky, and the greater would be its velocity in receding from us. These galaxies were, according to a famous analogy, like raisins in some giant loaf of raisin bread baking in the oven. As the loaf expands, each raisin moves apart from the others. If you were a raisin in such a loaf, then a distant raisin would move away from you more quickly than one nearby. Hubble had discovered that the whole universe was expanding —one of science's greatest results and biggest surprises.

Thus Einstein's universe model included a prediction that was proved false. In Einstein's model, galaxies would forever remain at the same distance from one another—they would not be moving apart.

Meanwhile, Alexander Friedmann had already found the answer. He solved Einstein's original equations exactly—without the cosmological constant—by making an important assumption: there were no "special" points in space. In other words, any point in space was as good as any other. That meant that as far as the curvature of space was concerned, no locations were special, and the amount of curvature must be the same everywhere.

The only thing left to specify was whether the curvature was positive (as on the surface of a sphere), zero (as on a flat tabletop), or negative (as on the surface of a western saddle). There were only three possibilities:

1. A *closed, positively curved three-sphere universe* with a spatial geometry like the one Einstein proposed. In this universe any triangle would have a sum of angles greater than 180 degrees. This universe is closed, with a finite circumference in all directions. It has a finite number of galaxies yet no boundary. Friedmann explored this case in 1922.

2. A *flat, zero-curvature universe*, where space is infinite in all directions and obeys the laws of Euclidean geometry—triangles always have a sum of angles equal to 180 degrees. This universe has an infinite number of galaxies. This intermediate case was added in 1929 by Howard P. Robertson of Princeton.

3. An *open, negatively curved universe* in which every triangle has a sum of angles less than 180 degrees. This negatively curved universe would also extend to infinity and have an infinite number of galaxies. Friedmann explored this case in 1924.

Friedmann then found that in the original version of Einstein's theory—with no cosmological constant—each of these models must be evolving, or changing, over time. The closed, three-sphere universe would have needed to start off with zero size. This was the moment of the big bang. Then it expanded, like the surface of an inflating balloon, to use an analogy developed by Sir Arthur Eddington. Galaxies would be like dots on this expanding balloon. As the balloon expanded, the dots would move apart, the distance between any two dots increasing with time. Eventually, the three-sphere universe would reach a maximum size and start to deflate, eventually shrinking back to zero size, causing a big crunch at the finish. Such a universe was finite in both space and time. The flat and negatively curved models also started with a big bang but expanded forever in the future. They were infinite in spatial extent and also infinite in time, in the direction of the future.

What does this mean for the motion of galaxies? Galaxies

attract one another by gravity, and yet they are moving apart at high velocity today. Do they have enough speed to escape from one another's gravitational pull and continue to move apart forever, or will their mutual gravitational attraction eventually overcome their outward velocities and make them fall together? If there is no cosmological constant, as Friedmann's models assume, the answer depends on the current density of the mass in the universe. If it is greater than a critical value, then the universe will eventually collapse, and the closed three-sphere, big bang–big crunch model applies. If the density exactly equals the critical density, then the flat model applies, and the universe will expand ever more slowly, just barely escaping collapse. If the density is less than this critical density, the negatively curved model applies, and the universe will continue to expand forever. Given the currently observed recession velocities of galaxies, this critical density is approximately 8×10^{-30} grams per cubic centimeter. That's equivalent to about 5 hydrogen atoms per cubic meter. According to Friedmann's models, if the average matter density of the universe today exceeds this critical value, the universe will collapse; otherwise, it will continue to expand forever.

Hubble found that the universe on large scales did indeed look the same in all directions, just as Friedmann's model implied. Clusters and groups of galaxies were sprinkled similarly wherever he looked. The counts of faint galaxies in different large regions of the sky were approximately the same. Furthermore, these groups and clusters of galaxies were all receding from us: the farther away they were, the faster they receded. This might look as if we were at the center of a finite explosion, but after Copernicus, we were not going to fall for that idea. Copernicus convincingly showed that Earth was not at the center of the universe, as people had previously thought. Even though our galaxy appeared to be at the center of a great explosion whose debris was receding from us equally in all directions, why should our galaxy be the lucky one at the center, making all the others off-center? If the universe looked the same to us in all directions, then it must look that way to observers on every galaxy—otherwise we would be special. The idea that our location is not special is called the *Copernican principle* and has been one of the most successful scientific hypotheses in the history of science. Hubble's observations that the universe appeared the same in all directions, coupled with the idea that we were not special, forced the conclusion that Friedmann's hypothesis must be true. If the universe looked the same in all directions as seen from all galaxies, then no special directions and no special locations existed. Friedmann's inspired guess now became a necessity, as pointed out by Howard P. Robertson of Princeton and Arthur G. Walker of Great Britain. And Friedmann's remarkable prediction that the universe should be either expanding or contracting was confirmed. (Unfortunately, he did not live to see this. Friedmann died in 1925, 4 years before Hubble announced his discovery.)

Friedmann's models meant that regardless of its curvature, the universe began in the finite past with a big bang. At the moment of the big bang there is a state of infinite density and infinite curvature—a singularity. This is a first cause.

The simplest of Friedmann's big bang models is the closed three-sphere universe that starts with a big bang and ends with a big crunch. Its spacetime geometry (shown in Figure 19) resembles a football. The big bang is the point at the bottom, and the big crunch is the point at the top. As drawn in the figure, time goes upward toward the future. The two-dimensional surface of the football shape shown in this diagram contains, for simplicity, one dimension of space (around the circumference) plus the dimension of time (from bottom to top). Ignore the inside of the football and the region outside. The surface of the football—the pigskin—is all that's real. To know what this universe looks

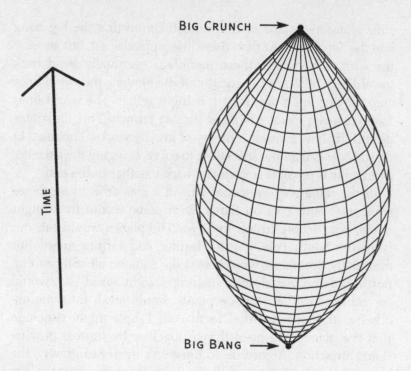

BIG CRUNCH →

TIME

BIG BANG →

Figure 19. Friedmann's Closed Three-Sphere Universe

like as a function of time, just cut the football along a horizontal section and move this cutting plane upward in time. The universe starts as a point (the big bang) and then becomes a growing circle. The circle represents the circumference of the three-sphere universe. The universe is expanding. When the "equator" of the football is reached, the circle reaches its maximum size. As the cutting plane moves higher, the circle gets smaller and smaller, finally ending as a point (the big crunch).

The galaxy world lines are geodesics, going as straight as possible along meridians of the football connecting the big bang and the big crunch. At first, these lines spread apart, but because the surface is curved, these meridians eventually bend back toward each other. At the equator of the football, the world lines stop moving apart and start moving together. The world lines finally converge and collide at the big crunch. This illustrates beautifully how Einstein's theory of gravity works. The mass in the galaxies causes the spacetime to curve, bringing the galaxies' as-straight-as-possible trajectories back together in the end.

In the same way, we could launch a squadron of airplanes from the South Pole on Earth. Each plane would fly straight north, not turning to left or right. The planes would fan out from the South Pole, getting farther and farther apart. But eventually, the planes would cross the equator, all still heading north, and find that, despite steering straight ahead, they would be coming together. All the planes would crash into one another as they reached the North Pole. People might conclude that the planes had been drawn together by mutual gravitational attraction. According to Einstein's theory of gravity, the world lines of galaxies are drawn back together because their mass curves spacetime.

Based on Friedmann's work, in 1948 George Gamow reasoned that the early universe right after the big bang must have been very dense and therefore very hot, just as when you pump up the air in your tire, cramming a lot of molecules into a small space, the molecules move around faster, and the air inside the tire heats up. Gamow deduced that the hot early universe would be filled with radiation that would cool as the universe expanded and became less dense. To picture this, think of the universe as an expanding circle, and consider a continuous wave of electromagnetic radiation going around the circle. It would look like a wavy circle, with a finite number of wave crests. These crests march around the circle as it expands. The number of wave crests does not change as the circle grows in size, so the wavelengths between crests become larger. Longer wavelength radiation has less energy and corresponds to a lower temperature. As the universe expands, therefore, the radiation in it loses energy, and its temperature drops.

Gamow also calculated the nuclear reactions that would occur as the universe expanded and cooled. After cooking at high temperature, the universe would emerge as mostly hydrogen (nucleus of 1 proton), about 24 to 25 percent by weight of helium (nucleus of 2 protons and 2 neutrons) and about 3 to 4 parts in 100,000 of deuterium by number (heavy hydrogen with a nucleus of 1 proton and 1 neutron). Tiny amounts of lithium would also be produced. Heavier elements such as carbon, nitrogen, oxygen, right up to uranium could be made later, after the big bang, in stars. Helium could be made in stars also. But there was no known way to make deuterium in stars. Nuclear reactions in stars burn up deuterium, making more helium. Gamow knew that small amounts of deuterium had been observed in the universe; the hot big bang seemed the only possible source of it. By knowing approximately how much deuterium now exists, Gamow could determine how much thermal radiation had been present at early times. He found that the deuterium we observe today was made only a few minutes after the big bang when the universe was a billion times smaller than it is today. Gamow's two colleagues Ralph Alpher and Robert Herman calculated what would happen to this radiation as the universe expanded to its present size. By the present epoch, they calculated, the radiation should have cooled to a temperature of about 5 degrees above absolute zero on the Kelvin scale. (Zero degrees Kelvin corresponds to −273 degrees Celsius or −459 degrees Fahrenheit.) This prediction, made in 1948, gave the radiation wavelengths in the millimeter range—microwaves.

At Princeton in the early 1960s, Robert Dicke independently concluded that after the big bang, the early universe must have been very hot. Gamow's paper had been forgotten. Dicke had a bright young colleague, Jim Peebles, calculate the nuclear reactions to find out how hot the universe should be at the current time, unwittingly repeating the Herman and Alpher calculations. Dicke, a master builder of microwave receivers, figured he could build a radio telescope capable of detecting the radiation, even though it would be of very low intensity. He, David Wilkinson, and P. G. Roll, also at Princeton, set about building the telescope, a horn-shaped contraption looking rather like a trumpet. Because the horn was pointed up toward the sky, hardly any contaminating radiation could leak in from Earth. Normal radio telescopes, having a big parabolic dish at the bottom and a receiver pointing down at it, are more susceptible to radiation contamination from Earth. Dicke thought he was building the only radio telescope in the world capable of detecting the thermal radiation left over from the big bang. He was wrong.

Only about 35 miles away at Bell Labs in Holmdel, New Jersey, Arno Penzias and Robert Wilson were already operating a larger horn antenna. It had been designed to receive microwave signals bounced off the newly launched, 100-foot-diameter Echo satellite, a thousand miles up in Earth orbit. To their surprise, Penzias and Wilson found microwave radiation was coming from all over the sky, corresponding to thermal radiation at a temperature of about 3 degrees above absolute zero on the Kelvin scale. This signal was unlike any other astronomical source. At first they thought the radiation might be coming from some pigeon droppings in the horn. But after a careful cleaning, they obtained the same results.

Penzias called up his friend, radio astronomer Bernie Burke, asking if Burke knew any astronomical sources that could produce 3-degree radiation equally from all over the sky. As it happened, Burke had just heard of a talk given by Jim Peebles, telling of the Princeton group's plan to look for this same radiation. He suggested that Penzias get in touch with Dicke. The

Princeton group was invited out to Bell Labs where, to their amazement, they saw the only other radio telescope in the world able to detect the radiation. They had been beaten. Penzias and Wilson and the Princeton group published side-by-side papers in the *Astrophysical Journal,* explaining the observations and the theory. It was 1965.

Five years later, I had the privilege of working with Penzias and Wilson on that same horn telescope. We were making some rather routine observations to calibrate the intensity of some known radio sources, but it was exciting for me nevertheless. I could see firsthand how careful they were and how well they worked as a team. Arno was the more ebullient of the two, Bob the quiet one. I remember Arno once calling out that a problem must exist with a particular electronics board; quick as a wink, Bob had pulled it out, tested a few junctions with a meter, replaced the offending part, and shoved the now-working board back in. This kind of fluid teamwork had led to their big discovery; they were careful enough to know that they had eliminated all other possible sources of contamination. Therefore the excess signal must have come from the sky.

I got to run the horn myself many nights, and I was thrilled as a graduate student to operate the telescope that had allowed people to see farther into space than any other telescope had. The thermal microwave background radiation it detected had last scattered off electrons and protons some 13 billion years ago, just 300,000 years after the big bang.

Working there, I once saw a letter from George Gamow pinned up on Penzias's wall. It congratulated Penzias on his recent review paper on the subject but complained that it left out some early history. Gamow pointed out that he had predicted the radiation in 1948 and that his colleagues Alpher and Herman had estimated its current temperature at 5 degrees. He gave the references.

The stationery on the letter said it came from the Gamow

dacha (the Russian term for a country house) in Boulder, Colorado. This brought me full circle. Growing up, I had read all of Gamow's books and was always a fan of his work on the hot big bang. My mother had a friend whose best friend was Gamow's wife, so when I happened to work at the Joint Institute for Laboratory Astrophysics in Boulder in the summer of 1967, she let the Gamows know of my presence. They were gracious enough to invite me to their home for dinner. Dr. Gamow drove to pick me up in his Rolls Royce. He was older than he looked on bookjacket photos but just as jaunty.

During dinner he posed entertaining brainteasers. In the basement he had an entire wall of his own books, which had been translated into many languages. He said how gratified he had been about Penzias and Wilson's discovery of the microwave background that he and his colleagues had predicted so long ago. Predicting that the radiation existed and then getting its temperature correct to within a factor of 2 was a remarkable accomplishment—rather like predicting that a flying saucer 50 feet in width would land on the White House lawn and then watching one 27 feet in width actually show up. One could call it the most remarkable scientific prediction ever to be verified empirically.

Penzias and Wilson's discovery, for which they received the Nobel Prize in physics, essentially clinched the case for the big bang model. Three decades later the Cosmic Background Explorer (COBE) satellite measured this cosmic microwave background radiation at many wavelengths to exquisite precision, finding its temperature to be 2.726 degrees above absolute zero on the Kelvin scale. These observations were so dramatically in accord with the thermal radiation predicted by Gamow, Herman, and Alpher that the audience of physicists and astronomers gathered in Princeton to hear the COBE results in 1992 burst into spontaneous applause when David Wilkinson put the COBE spectrum slide on the screen.

COBE later detected small fluctuations in the temperature—1 part in 100,000—based on its observations of different directions in the sky. Such small fluctuations in the radiation and matter density present in the early universe resemble tiny ripples in a quiet pond, but they can grow into crashing waves later. Regions of slightly above average density gravitate more strongly than surrounding regions and gather even more matter onto themselves. By this process, the fluctuations in the density in the early universe implied by our observations of the microwave background radiation can develop into the galaxies and clusters of galaxies we see at the present time.

With the big bang model in ascendancy, attention focused on the big bang singularity itself. Stephen Hawking and Roger Penrose proved some theorems showing that, barring quantum gravity effects and closed timelike curves, if the energy density in the universe is always positive and the pressure is never negative enough to produce a net repulsive gravitational effect, then the level of uniform expansion we observe today implies that an initial singularity had to occur. In other words, initial singularities would form even in models that are not exactly uniform. This initial singularity was taken to be the first cause of the universe. But this conclusion invited questions concerning what caused the singularity and what happened before it. The standard answer for what happened before the big bang singularity is this: time was created at the singularity (at the bottom of the football in Figure 19), along with space. Thus, time did not exist before the big bang, and thus nothing happened before it. Asking what happened before the big bang is like asking what is south of the South Pole. A neat answer.

But a troublesome question remains: what caused the initial singularity to have almost, although not perfect, uniformity, so that the microwave background radiation doesn't exhibit vastly different temperatures in different regions of the sky?

Another problem is that singularities are usually "smeared"

by quantum effects. Heisenberg's uncertainty principle tells us that things cannot be located exactly; it's as if you took a pen and made a dot but then erased it to smear the ink over a region of the page. Such quantum fuzziness may stop the density from reaching an infinite value. As we trace time back toward the initial singularity, following the laws of Einstein's theory of general relativity, we first reach an epoch in which the density becomes so large that quantum effects cause the laws of general relativity to break down. At this density (5×10^{93} grams per cubic centimeter), quantum uncertainties in the geometry of spacetime become important; spacetime is no longer smooth but instead becomes a complicated spongelike spacetime-foam. Thus, we cannot retrace our way confidently back to a state of infinite density; we can only say that we would eventually reach a place where quantum effects should become important and where classical general relativity (assumed by Hawking and Penrose in deriving their theorems) no longer applied. We do not currently have a theory of quantum gravity or an all-encompassing theory-of-everything (unifying gravity, the weak and strong nuclear forces, electromagnetism, and quantum mechanics) to help us. Instead, we have to admit that before a certain time, we do not know what happened—much as geographers of old had to mark *Terra Incognita* on maps. We cannot say exactly how our universe formed.

An Oscillating Universe

Some physicists in the 1960s speculated that quantum effects might allow a cosmology collapsing toward a big crunch to "bounce" and make another big bang. This could lead to an oscillating universe, with an endless sequence of big bangs and big crunches. The oscillating model avoids the first-cause problem by the "it's turtles all the way down" solution recounted by Carl Sagan in *Broca's Brain* in a chapter entitled "Gott and the

Turtles." This chapter described some work I had done with my colleagues Jim Gunn, Beatrice Tinsley, and David Schramm, which suggested the universe would continue to expand forever rather than bouncing. In this chapter Sagan told the story of a traveler in olden days who, encountering a great philosopher, asked him to "describe the nature of the world."

"It is a great ball resting on the back of the world turtle."

"Ah yes, but what does the world turtle stand on?"

"On the back of a still larger turtle."

"Yes, but what does he stand on?"

"A very perceptive question. But it's no use, mister, it's turtles all the way down."

In the oscillating model, therefore, the answer to what caused our universe is "the collapse of the previous universe." What caused that universe? Well, the universe before that—and don't worry, it's universes all the way down. In this model, an infinite number of expansion and contraction cycles make up the Universe (note here I am using a capital letter U—this will denote the ensemble of causally connected universes sometimes called the *multiverse*, or as Timothy Ferris says, the "whole shebang"). The Universe then consists of an infinite number of closed big bang models laid out in time like pearls on a string (see Figure 20). There is no first cause because the Universe has existed infinitely far back in the past. The Universe (the infinite strand of pearls) has always been and will always be in existence, even though our pearl, the cycle containing our standard closed big bang cosmology, has a finite duration. The model brings us back both to Aristotle's eternal universe and close to Einstein's original conception of a closed universe with infinite duration to the past and future, although this version oscillates rather than remaining static.

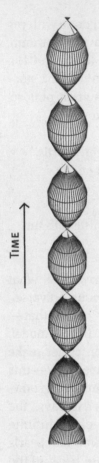

TIME →

Figure 20.
Oscillating Universe

The oscillating Universe was thought to have some problems with entropy, the scientific name for disorder. Break a vase, and the pieces fly in different directions —the disorder in the Universe increases. Place an ice cube on the stove—it melts, and disorder grows again. The solid ice cube, with its molecules held in regular locations, is more ordered than the chaotic placement of the molecules in the liquid state. On Earth, we sometimes see order rising locally, as when we make ice cubes in the freezer. But that takes energy. When we burn fuel in a power plant to produce this energy, that process leaves the plant itself more disordered than before, outweighing the extra order produced in the ice cubes in the freezer. If we carefully tote up the total disorder in the universe, we see that it is increasing with time (this is the second law of thermodynamics).

At late times (far along in the history of a given universe), we expect that particular universe to be very irregular and chaotic as it collapses to form the big crunch. Since entropy increases with time, how could this disordered, or high-entropy, state recycle into the highly ordered, low-entropy, nearly uniform state of the next big bang? One might hope to occasionally obtain by chance a big bang as nearly uniform as the one that produced our universe, like throwing baskets of coins up forever and being lucky enough, once in a great while, to have all the coins land heads up. But the highly ordered region of the universe we see is very large—with a radius of 13 billion light-years. Universes displaying smaller regions of order would be much more common. Random astronomers in such an oscillating Universe might not be expected to see initial conditions in their big bang as uniform over as large a region as we observe in ours. Thus, the widespread, nearly perfect uniformity of the initial conditions that we observe in our universe still remained a mystery throughout the 1960s and 1970s.

Inflation

Alan Guth's 1981 theory of inflation offered an explanation for why the initial conditions in the big bang should be approximately but not exactly uniform.

We observe today four fundamental forces in our universe: the strong and weak nuclear forces, electromagnetism, and gravity. These forces have different strengths, gravity being the weakest. In the very early universe these forces may have been equal in strength and united in one force that may someday be explained by the hoped-for theory-of-everything. Thus, in the early universe, the laws of physics would be different. Therefore, the cosmological constant—the energy density of the vacuum—could also be different in that early phase. Einstein didn't consider this possibility. Guth proposed that in the very early universe, the energy density of the vacuum would be enormous and constitute the dominant form of energy in the universe; it would therefore shape its geometry according to Einstein's theory of general relativity. What would this geometry look like? That answer was already known.

When Einstein thought of the cosmological constant in 1917, he used it plus ordinary matter to produce a static cosmology. But later in that same year Dutch astronomer Willem de Sitter wondered what would happen if a universe had a cosmological constant and nothing else. The result is called *de Sitter spacetime*, illustrated in Figure 21. To understand inflation, one must understand de Sitter spacetime, which looks like the surface of an infinite hourglass—one cone to the past, and one cone to the future, joined by a narrow waist. Like previous cosmology diagrams, this one illustrates only one dimension of space, wrapped around horizontally, and the dimension of time, which is shown vertically. This is a closed three-sphere universe starting off in the infinite past with infinite size, contracting at nearly the speed of light. The repulsive effect of the cosmological constant causes this contraction to slow down and then reverse. The universe reaches a minimum size, then starts to expand, slowly at first but then faster and faster, eventually approaching the velocity of light. Cut the model in Figure 21 with a horizontal plane, and the cross section forms a circle, showing the circumference of the three-sphere universe at that time. Move the horizontal plane slowly from the bottom to the top of the diagram and you will see the circular cross section shrinking at first, reaching minimum size at the narrow waist, and then expanding.

If there were particles in such a universe, they could have world lines as indicated by the vertical curved lines. These lines approach one another at first, draw closest together at the waist, and then, like corset stays, fan out at the top. As the particles increase in speed and approach the speed of light, the de Sitter spacetime begins to resemble a cone angling outward by 45 degrees. As the particles begin to move outward at speeds approaching the speed of light, their clocks begin to tick ever more slowly, according to special relativity. As the ticks spread farther and farther apart, the universe expands more and more between ticks. In fact, the particles "see" the circumference of the universe growing exponentially (2, 4, 8, 16, 32, 64, and so on) as a function of their own clock time.

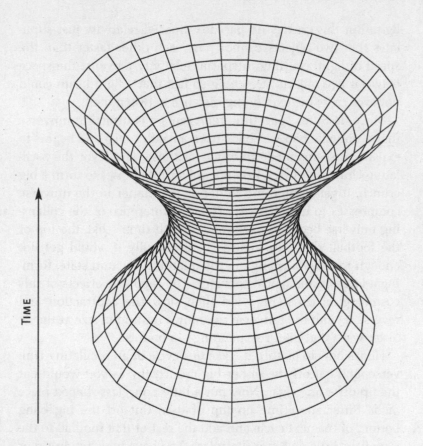

TIME

Figure 21. de Sitter Spacetime

Guth called the de Sitter expansion phase of the universe "inflation" because the size of the universe kept doubling—like prices in a period of severe monetary inflation. The distance between two particles would grow exponentially as well, as measured by their own ever-slowing clocks. Eventually, they would think they were moving apart faster than the speed of light. But this creates no paradox. Special relativity just stipulates that two objects cannot *pass* each other faster than the speed of light. Nothing in special relativity prevents the space between two objects stretching so fast that a light beam could not ever cross the expanding distance between them.

Since de Sitter spacetime represents a contracting universe that eventually bounces back from a near crunch and begins to expand, it seems made to order for those who favor the oscillating Universe model. When a universe collapses to form a big crunch, it gets hotter and hotter as the matter in the universe compresses to high density. The circumference of the collapsing universe becomes ever smaller with time—like the top of the football shape in Figure 19. Eventually it would get hot enough to cause a change in the quantum vacuum state, forming a large cosmological constant. The repulsive effects of this cosmological constant could then slow the contraction and reverse it, with the universe reaching a minimum size at the de Sitter waist and then reexpanding—a bounce.

Here's how to picture it. Take one cycle of an oscillating Universe, and cut off the upper big crunch tip, as you would cut the tip off a fine cigar. Now put a little hourglass-shaped piece of de Sitter spacetime on top of that. Cut off the big bang bottom of the next cycle and add the rest of that football to the top of the little de Sitter hourglass. You have now gotten rid of the big bang and big crunch singularities and bridged the gap between the collapse phase of one football-shaped oscillation cycle and the next by an hourglass-shaped piece of de Sitter spacetime. The circumference of de Sitter spacetime at its waist can, in this scenario, be very small indeed, 10^{-33} to 10^{-26} centimeters. That almost resembles a point—almost a singular big bang. To an observer in the expanding phase later, this universe would look just like a big bang model.

According to Guth, a de Sitter phase answers the question of how the universe's expansion got started: it was the repulsive gravitational effects of the early cosmological constant that started the expansion. Eventually, the high-density inflationary vacuum state decayed into the normal vacuum. Guth figured that, at this point, the energy density in the inflationary vacuum became converted into normal, hot thermal radiation, and the expansion proceeded just as in the big bang model. With only normal radiation and matter present, the expansion could now slow with time, just as in the big bang model.

Guth now understood why the big bang would be so uniform. Regions that have had time to trade light signals should equilibrate to the same temperature. Then as the universe inflates, doubling and redoubling in size, these regions effectively pass out of causal contact as they are no longer able to trade light signals. But after the inflationary vacuum decays, the expansion slows and these regions come into contact again. As astrophysicist Bill Press has noted, they say hello, goodbye, and then hello again. When regions that have reached the same temperature before parting say hello again, they are also in thermal equilibrium. When we look out at the microwave background in different directions, we see stuff that is all at approximately the same temperature. These different regions were originally close enough to one another to have exchanged photons in the early de Sitter phase of the universe. This kind of equilibration would not be possible in the standard big bang model, where the regions we see today would be "saying hello" to one another for the first time now. Thus, if our universe started with a high-density vacuum state instead of a big bang singularity, it could explain why the microwave background has the high degree of uniformity that we observe.

But it's not perfectly uniform. Regions originally in causal contact were so small that the uncertainty principle demands that they should have appreciable quantum fluctuations in energy density from place to place. As James Bardeen of the University of Washington and his colleagues Paul Steinhardt and Michael Turner have shown, these fluctuations would be frozen when the regions passed out of causal contact and would have approximately the same magnitude (approximately 1 part in 100,000) when they "said hello" again. But these regions would inflate vastly in size while out of causal contact and while the universe kept doubling in size.

According to the inflationary model, a region originally 10^{-26} centimeters or smaller in width could grow to be billions of light-years in expanse. We can calculate how fluctuations would evolve in an inflationary universe and compare these calculations with our observations of the microwave background. The results agree very well.

An important point is that the quantum fluctuations in density predicted by inflation should be random. Therefore, the three-dimensional geometry of the high- and low-density regions forming in the universe should be equivalent. This is possible with a *spongelike* geometry, as pointed out by Adrian Melott, Mark Dickinson, and me in 1986, and further developed by Andrew Hamilton, David Weinberg, Changbom Park, Michael Vogeley, Trinh Thuan, Wes Colley, me, and other colleagues. A sponge has insides and outsides that are similar in shape. A variety of galaxy samples now measured by many groups all show a spongelike distribution of galaxies. The largest sample, including over 15,000 galaxies, shows excellent agreement with the theory. It's remarkable that the structures we see in our universe today may well be the fossilized remains of quantum fluctuations occurring during the first 10^{-35} seconds of our universe.

Inflation also explains why our universe is so large. It just keeps doubling and redoubling in size. Such a sequence (2, 4, 8, 16, 32, 64, and so on) grows quickly: just 10 doublings, and the universe is a factor of a thousand larger; 20 doublings, and the universe is a factor of a million larger; 30 doublings, and it's a billion times larger. The universe may have undergone more than 100 doublings, increasing its size by a factor of more than 10^{30}, during the inflationary epoch.

Curiously, that cosmological constant Einstein invented—his biggest blunder, he thought—has now come to the rescue in the form of an inflationary vacuum to explain the early universe.

An inflationary epoch could provide the bounce that turns a big crunch into a big bang. But what if one could just start off, say, with the narrow waist of de Sitter spacetime, thereby eliminating the contracting de Sitter phase altogether as well as any previous universes? The inflationary universe at its waist is very small; it is a closed universe with a tiny volume—much smaller than that of a proton. Yet it starts expanding, eventually giving rise to the enormous universe we see today. Indeed, Guth noted that one could start with any tiny bit of inflationary vacuum state and, as the expansion took place, an ever larger volume of inflationary vacuum would be produced. You can't start with nothing, but you can start with something really tiny.

Start with any tiny bit of inflationary vacuum state, and it will grow endlessly. In fact, that was the only problem with Guth's original paper. As Guth himself noted, getting a graceful exit from inflation was problematic. Still, because the inflationary vacuum state was of positive energy density, it was ultimately vulnerable to decaying into the lower-energy-density normal vacuum.

According to Guth, inflation might end when the energy in the inflationary vacuum was dumped in the form of thermal radiation over the whole space at once. That would be like boiling a kettle of water on the stove, only to find that the whole kettleful had suddenly turned into steam. The distribution of steam would be uniform like the uniform hot big bang model we see, but this is unlikely to happen. As you know, when you boil water on the stove, bubbles of steam form. Indeed, Sidney Coleman of Harvard and his colleague F. de Luccia had shown that a sea of high-density vacuum would likely decay by forming bubbles of ordinary vacuum within it. Each bubble would expand after its formation, its wall eventually traveling outward at nearly the speed of light. The vacuum state inside a bubble is a normal vacuum with zero energy density and a zero pressure. Outside the bubble, the pressure would be negative (a universal suction), so the inflationary vacuum state outside simply would pull the bubble wall outward, making it expand. Nevertheless, the bubbles would never percolate to fill the entire space. Two bubbles born near each other would collide as they expand, but two bubbles born far apart would never be able to expand fast enough to close the gap between them, because the gap itself is stretching so rapidly. The result is an ever-expanding, high-density vacuum sea containing isolated bubble clusters. This is a non-uniform distribution, quite unlike the uniform universe we see. Was Guth's wonderful and powerful theory actually stillborn? No. Guth had noted the problem. The solution required a closer look at those bubbles.

Bubble Universes

Coleman and de Luccia had shown that when a bubble forms by a quantum process in a high-density inflationary vacuum, the bubble wall starts at a nonzero size. Then the wall expands outward, faster and faster (see Figure 22). Light beams emitted from the center of the bubble at its creation would never quite catch up with the outward-rushing bubble wall, which has a head start and draws ever closer to the speed of light.

Coleman and de Luccia thought the entire inside of the bubble would be empty—flat spacetime filled with nothing but the normal vacuum. But I thought that if inflation could continue inside the bubble, one could form an entire inflationary universe like ours inside just one bubble.

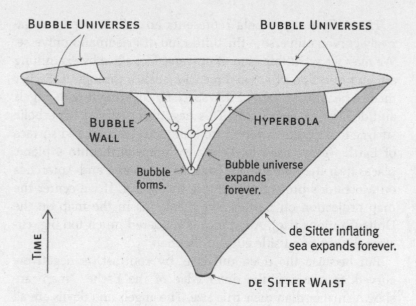

Figure 22. Bubble Universes Forming in a High-Density Inflationary Vacuum

Imagine rocket ships passing through a bubble-initiating event, *E*, at different speeds. These rocket ships then fan out in all directions. Assume that their clocks were synchronized at noon, when they were all together at *E*. Then let their alarm clocks all go off at one o'clock. This is illustrated in Figure 22. The event *E* is shown by an alarm clock at noon. Three observers' world lines crossing through event *E* are shown as three lines with arrows extending upward. The three alarm clocks are shown going off at one o'clock. The hyperbola shows the surface in spacetime where the alarm clocks all go off; it curves upward toward the future as it spreads outward because clocks on rockets moving rapidly to the left or right tick more slowly, according to special relativity, and take longer to reach one o'clock.

This infinite hyperbola represents an infinite, open, negatively curved universe—the third kind of Friedmann universe. We may see what this kind of universe looks like by examining a map projection of a two-dimensional slice through it. Escher made a beautiful "map" of this space by covering it with angels and devils (see Figure 23). This negatively curved hyperbolic surface looks quite different from the positively curved surface of Earth we are used to. When we map Earth onto a plane, places near the map's edges (such as Greenland and Antarctica on Mercator's projection) are drawn too big. If we center the map projection on Earth's North Pole (as in the map on the United Nations flag), Antarctica is rendered much too big, circling the entire outside edge of the map.

But because the open universe, by contrast, is negatively curved, things near the outer edge of the Escher "map" are shown smaller than their true size. The angels and devils are all really the same size. Count out several angels and devils from the center, and then trace around a circle at that radius. You will see hundreds of angels and devils crowding along that circumference. In this kind of space, the circumference of a circle is bigger than one would expect from Euclidean geometry. Getting lost in an open universe would be easy. Each angel and devil in the diagram represents a triangle with angles of 60 degrees at the feet, 45 degrees at the left wingtip, and 45 degrees at the right wingtip. (Six angels and devils come together at a point at their feet, and a 360-degree circle around that point divided into 6 equal parts gives a 60-degree angle at each pair of feet. Similarly, 8 angels and devils meet at a point where their wingtips touch, giving a 45-degree angle for each wingtip—360 degrees divided by 8.) Each triangle has 3 angles—of 45, 45, and 60 degrees—which add up to 150 degrees, less than the 180 degrees expected in Euclidean geometry, confirming that this is a negatively curved space. There are an infinite number of angels

Figure 23. *Circle Limit IV* (1960), by M. C. Escher.
This image shows a negatively curved, open universe.

and devils, stretching to infinity, and each line through the center of the map represents an infinitely long hyperbola, like the one shown in Figure 22.

If the bubble is empty, one gets an empty open universe of zero density, as Coleman and de Luccia noted. But, as I pointed out, if the energy density in the inflationary vacuum state continues to be high until it is dumped in the form of thermal radiation at one o'clock, according to the alarm clocks in Figure 22, this transition will occur on a hyperbolic section, creating an open Friedmann universe, infinite in extent (with an infinite number of galaxies) and expanding forever. Each of the three observers in Figure 22 will think he or she is at rest at the center, directly to the future of the event E (the alarm clock reading noon), and that the others are expanding away from him or her. Likewise, each person on Earth thinks correctly that the center of Earth lies directly beneath him or her. Just as there is no real center of Earth to be found on Earth's surface, there is no center of the universe to be found in the universe today. Just as the real center of Earth lies below us, the real center of our universe (the event E) lies in our past.

The whole expanding universe, with world lines fanning out from E at speeds slower than the speed of light, could fit inside the ever-expanding bubble wall. Interestingly then, a whole open inflationary universe could sit inside one of Coleman's bubbles. I said in my research paper that our universe was just one of the bubbles. I thought this could address Guth's problem. From a viewpoint inside one of the bubbles, what you see is uniform: our bubble is uniform. We do not see any other bubbles because when we look out, we are looking back in time and so see our own bubble and the inflationary sea that preceded it. No other bubbles have collided with ours yet.

In other words, bubbles weren't the problem—they were the answer.

My paper on open bubble universes was published on January 28, 1982, in *Nature*. Later, its key diagram would be chosen for the cover of the annual called *Physics News in 1982*, published by the American Institute of Physics. For a physicist like me, this felt like getting my picture on the cover of *Rolling Stone*. My paper even found its way into the reference list in the *Star Trek* novel *The Wounded Sky* by Diane Duane—along with Mr. Spock's famous paper "Mathematical Implications of

Nonhomogeneous Paratopological Convergences Between Orthogonal Unbridged n-Spaces, with Substantiating Field Measurements" from the *Review of Modern Hypercosmology and Cosmogony*, Vol. 388, Stardate 9258.0, and a paper by the noted Vulcan physicists T'pask, Sivek, B'tk'r, and K't'lk, from the *Proceedings of the Vulcan Science Academy*. I was quite honored to appear in such company!

My paper stated that inflation had to continue for a while inside the bubble, but I had no good mechanism for accomplishing this. On February 4 and April 26 of that year, independent papers by Russian physicist Andrei Linde and by Andreas Albrecht and Paul Steinhardt, working at the University of Pennsylvania, were published, giving detailed particle-physics scenarios that produced just such a model. They used the idea that the quantum vacuum could have different energy densities at different locations in space and time. In the same way, a landscape may have different altitudes at different locations. (See Figure 24.) The normal vacuum with zero density would be at sea level. A high-density inflationary vacuum would be represented by a point in a high mountain valley. A bowling ball sitting on a mountain would roll down to sea level, releasing some energy as it fell. But if the ball is in a mountain valley, surrounded by peaks on all sides, it could not roll down. According to Guth's inflationary model, when the universe began, it was trapped in such a high-density vacuum state. As long as it remained trapped in a mountain valley, it continued to inflate. It would have remained trapped forever but for the effects of quantum mechanics. Quantum mechanics allows a finite probability that a bowling ball can simply tunnel through the surrounding mountain ridge and emerge on a slope from which it can eventually roll down to sea level (see Figure 24).

Quantum tunneling has been observed. When uranium decays, it spits out a helium nucleus, which shoots outward because of electrostatic repulsion. From the energy of the escaping

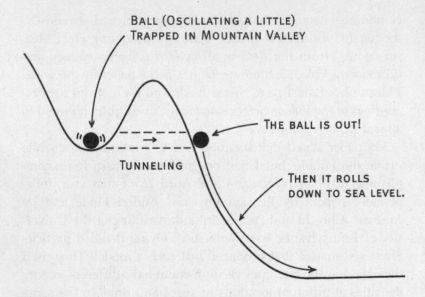

Figure 24. Quantum Tunneling

helium nucleus we can tell how close it was to the original uranium nucleus when it was emitted. Surprisingly, we find that its trip had to begin well *outside* the nucleus. Had it escaped from the known outer edge of the nucleus, it would have shown a much higher level of energy from being repelled much more forcefully. How did it suddenly get so far outside? (This question calls to mind an ancient Zen koan: "How did the duck get out of the bottle?" One answer: "The duck is out!") The helium nucleus simply tunnels out of the uranium nucleus, suddenly appearing outside without ever passing through the space in between. George Gamow—later of big bang fame—figured this out in 1928.

Tunneling is what happens when a Coleman bubble forms. The bowling ball trapped in the valley represents the original high-density vacuum state. The bowling ball tunneling out of the

valley represents the formation of a bubble centered on an event
E. The bubble suddenly appears, already of nonzero size—the
bubble is out. If the bowling ball dropped immediately to sea
level, that would leave the bubble empty, with only a normal
vacuum inside. A big breakthrough, made by Linde and by
Albrecht and Steinhardt, was the idea that once the bowling ball
tunneled out, it would emerge on a high plateau. Here it would
roll along for a while before plunging down a cliff to sea level.
On the high plateau, where the vacuum state had high density,
inflation continued inside the bubble. Falling off the cliff occurred a certain amount of time after the event E, along a hyperbolic surface (where the alarm clocks read "one o'clock" in
Figure 22). This drop-off released energy, produced radiation,
and turned the expanding bubble into a hot open big bang
model. Neither Linde nor Albrecht and Steinhardt noted that
the bubble model made an open universe. They just figured that
the period of inflation inside the bubble would be long enough
to produce a model that would be nearly flat today. Guth noted
that if enough inflation took place, regardless of what shape the
universe took originally, it would look flat if it were large
enough today. (For example, take an elephant, and blow it up a
trillion times in size. Any little piece of it will look flat.) Guth
especially liked this feature of inflation, noting that if we observe the universe to be approximately flat today, then a lot of
inflation might easily explain how it got that way.

Current data on the microwave background seem to favor a
model that is approximately flat today. This doesn't preclude an
open universe but merely indicates that the inflation within the
bubble must have gone on for a long time. Then the entire 13-
billion-light-year radius out to the microwave background radiation that we can see currently might fit inside the big toe of
one of Escher's angels. The tiny part of the universe we can see
would then look approximately flat (just as the Bonneville Salt
Flats appear approximately flat even though they are really a

tiny portion of Earth's curved surface). The smoking gun proving that a bubble existed originally would be "forgotten" as the universe grew so large that the negative curvature inherited from the bubble's formation became undetectable. In this case, the universe could have formed in various ways; not only a bubble but an inflating region of some other shape could have been responsible.

Numerous lines of evidence suggest that the density of matter in the universe today (including the presumed dark matter holding together clusters of galaxies) is significantly less than the critical density required to produce a nearly flat universe. The best chance for additional density comes from having a tiny residual vacuum energy density today of about 6×10^{-30} grams per cubic centimeter—a small cosmological constant. Recent measurements of the recession velocities of distant supernovae by Saul Perlmutter of Berkeley, Robert Kirshner and Adam Reiss of Harvard, and their colleagues have supported this view by showing that the universe's expansion seems to be accelerating—like de Sitter space at late times. If that turns out to be the case, we have a small cosmological constant today, just as originally proposed by Einstein! The old boy would have been right after all—but for different reasons. Another happy person (if he were still alive) would be Abbé Georges Lemaître, who early in the 1930s proposed a cosmology that began with a big bang but ended with an accelerating expansion due to a small cosmological constant. In any case, the current data suggest that the universe will continue to expand forever.

A Universe from Nothing?

The bubble universe idea works only if the inflating universe had a beginning. If inflation extended infinitely into the past, its geometry would assume that hourglass shape shown in Figure 21. Bubbles forming in the infinite contracting phase that preceded the waist would run into each other. Like blow-fish puffing up in a shrinking pond, they would soon fill the space entirely. A. Borde and Alexander Vilenkin of Tufts University showed in 1994 that this would cause the inflationary vacuum state to decay into a contracting froth of bubbles that would end with a big crunch before the waist could be reached and before any reexpansion could occur. Start at the waist, however, as shown in Figure 22, and one can make an infinite number of bubble universes. Although each bubble grows ever larger, the space grows faster still, making room for more and more bubbles.

Vilenkin had an idea for how to begin the inflating universe at the waist: by using that peculiar feature of quantum mechanics we have already encountered—tunneling. Something peculiar might be needed to jump-start the universe; maybe it could have been tunneling.

Imagine a bowler rolling a bowling ball up a mountainside. The ball would roll up the mountain for a while, pause momentarily at its highest point, and then begin rolling back down toward the bowler. Likewise, although an infinitely old de Sitter space at first shrinks rapidly, it slows down, stops for just a moment at a minimum radius (at the waist), and then reexpands. But suppose we have a bowling ball sitting in some mountain valley. Eventually, by quantum mechanics, it tunnels through the mountain range and emerges on the mountainside, whereupon it starts rolling down the mountain. Such a tunneling process, whose probabilities were worked out by Vilenkin and Linde, could explain how the de Sitter universe started at the waist and then expanded outward.

But the ball existed somewhere before it tunneled—it was sitting at the bottom of the mountain valley. This state corresponds in this case to a closed universe of zero size (the point at the bottom of the black region of Figure 25). It isn't quite nothing, but it's as close to nothing as one could get. During the quantum tunneling through the mountain range, the geometry (the black region in Figure 25) can be described as a curved four-dimensional surface having four dimensions of space and no dimension of time. (The squares of distances in all directions are positive.) At the very bottom, the space starts off as a circle of zero size, a point, like Earth's South Pole, and keeps expanding as circles of latitude do on Earth, until they reach the equator, which represents the epoch in which the tunneling is complete and the Universe emerges from the "mountainside" to become the waist of a de Sitter spacetime. The Universe then expands outward toward the future (where it could eventually spawn an infinite number of bubble universes [as in Figure 22]). This makes a Universe that looks rather like a badminton shuttlecock. The white portion has three spatial dimensions and one time dimension. As H. G. Wells's Time Traveler might say, the black portion is a four-dimensional space that lasts for no time at all. No clocks tick in the black portion. It is a frozen, pure piece of geometry, bounded on the bottom by a point and on the top by a three-sphere (a circle in the figure) where it joins the waist. Normal time begins at the waist.

Hawking and Hartle have pointed out that the origin of the Universe can be traced back, in this case, to a south pole. They have noted that this pole is no different in kind from the other points in the black portion. They have pointed out that this tunneling geometry fulfills what they call a "no-boundary condition" that eliminates the initial conditions, such as the initial big bang singularity. The Universe might then be said to have provided its own initial conditions; in other words, it just *was*. The beginning had no loose ends. The black tunneling surface has an ending boundary (that joins it to the waist) but no starting boundary. (In the same way, Earth's Southern Hemisphere could be said to be bounded only by the equator at the top,

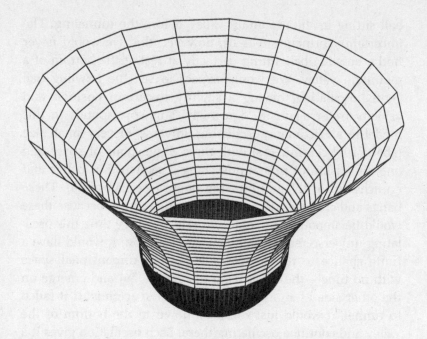

Figure 25. A Universe That Has Tunneled from Nothing

with no bottom boundary.) The idea harks back to one of E. P. Tyron's in 1973—that our universe might have formed spontaneously from a quantum fluctuation.

But to me a problem exists with this model. It doesn't really start with nothing; it starts with something—a quantum state, which tunnels out to become a normal spacetime.

Now back to Li-Xin Li. He found a further possible problem with this model. Quantum tunneling usually has *two* ends, for, as the word implies, a tunnel has two ends. On both ends there is an allowed spacetime. The tunneling universe can also be interpreted this way—as really starting from an oscillating Friedmann universe of zero size, corresponding to the bowling ball sitting in the mountain valley before the tunneling. The uncertainty principle tells us, however, that we would never find a bowling ball sitting perfectly at rest at the bottom of a mountain valley. We would expect to see the bowling ball gently oscillating, having a slightly uncertain position and velocity (refer again to Figure 24). This corresponds in this model to a Friedmann oscillating universe, of very small size (about 10^{-33} centimeters) with three dimensions of space and one dimension of time, undergoing a sequence of bangs and crunches prior to the tunneling phase (see Figure 26). These bangs and crunches do not produce singularities because these would be smoothed by quantum effects. Every time this oscillating universe reached maximum expansion, it would have a finite chance to tunnel (becoming a four-dimensional space with no time—the black region in Figure 26) and emerge on the other side as an expanding de Sitter spacetime. If it failed to tunnel, it would just roll back down to the bottom of the valley and continue oscillating there. Each oscillation gives it a chance to tunnel out, so eventually it would. In Figure 26 the tunneling (in black) connects two ordinary spacetimes (both unshaded), linking an oscillating universe to an inflationary one. The tunnel has two ends.

But again, where did the original oscillating universe come from? It couldn't have been around forever because, like a radioactive nucleus, it has a finite lifetime. Explaining its origin is a problem, putting us right back where we started.

Vilenkin still thinks that we can get along without an oscillating precursor universe. He likens the expanding de Sitter space to an expanding bubble wall in a still higher dimensional inflating space. Before the bubble forms, there is no oscillating wall solution, he argues. But there is *something* before—in this case, a higher-dimensional inflating sea. Li-Xin Li and I both favor the requirement that a quantum tunneling event have two ends, connecting two ordinary spacetimes. If we say some-

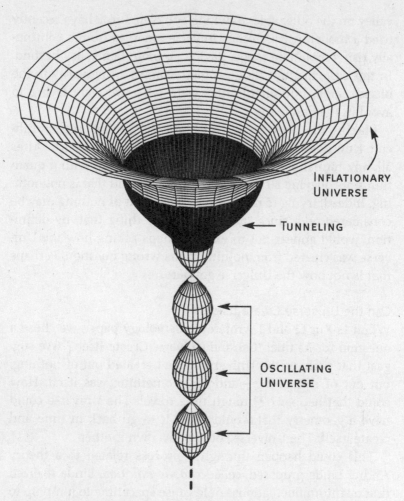

INFLATIONARY
UNIVERSE

TUNNELING

OSCILLATING
UNIVERSE

Figure 26. Eventual Tunneling of a Small Oscillating Universe

thing tunnels out, then it must have been something before that occurs. Indeed, the tunneling phase of Hartle and Hawking's model showed no singularities in its geometry (in the black region in Figure 25) precisely because there *is* a mountain valley on the other side. Hawking and Neil Turok have recently tried a model in which the Universe emerges from a continually rising mountain range, having no mountain valley beyond. In this case, the tunneling geometry shows a singularity in the black region, but singularities were what we were trying to avoid in the first place.

Making the Universe out of literally "nothing" seems difficult. How does "nothing" know about the laws of physics? After all, any tunneling-from-nothing model starts out with a quantum state obeying all the laws of physics—and that is not nothing. Indeed, trying to make the Universe out of nothing may be considered odd, since "nothing" is something that, by definition, would appear not to exist. Perhaps asking how the Universe was created from nothing is the wrong question. Perhaps that is not how the Universe got here.

Can the Universe Create Itself?

When Li-Xin Li and I wrote our cosmology paper, we chose a question for its title: "Can the Universe Create Itself?" We suggest that perhaps the Universe wasn't created out of nothing, but out of something—and that something was itself. How could that happen? Through time travel. The Universe could have a geometry that would allow it to go back in time and create itself. The Universe could be its own mother.

This could happen through a process related to a theory Andrei Linde proposed, called *chaotic inflation*. Linde realized that quantum fluctuations could cause spacetime to jump up to a higher vacuum energy density and a higher rate of inflation. (Imagine a bowling ball suddenly quantum-jumping up from the coastal plain into the mountains.) This explained how inflation could arise under very general circumstances, and it has become the standard inflationary scenario discussed today. According to Linde, because of these quantum fluctuations and their jumps in the rate of inflation, an inflating universe could sprout baby universes in the way that branches grow from a tree trunk. Each baby universe would then inflate to a size as large as the "trunk," and would bud its own baby universes. This would continue forever, with inflating universes continually branching off older branches, making an enormous fractal tree. (For additional details on this important theory, see the Notes.) Unless you are willing to account for this process by accepting the "it's turtles all the way down" solution, logic requires you to ask how the "trunk" arose.

In our cosmology paper, Li-Xin Li and I proposed that one branch simply curved back around to become the trunk. Figure 27 shows four inflating baby universes from left to right. At late times each of these expanding "trumpets" is an inflating de Sitter spacetime. Since each has a beginning (a waist) at the branch point, each of these can spawn an infinite number of bubble universes like those depicted in Figure 22. Once again, the surface is what counts in this diagram. Each trumpet can expand forever without encountering the others. The universes on the far left and right have not given birth to any baby universes yet, but given enough time, they will. Each baby universe has been created by the same branching mechanism. The laws of physics apply everywhere, and there are no singularities. As for the odd loop at the bottom, well—that comes from a baby universe that has looped back in time to become the trunk.

Admittedly, the geometry looks quirky. My wife noted that it resembles one of Dr. Seuss's fanciful illustrations. Neil de Grasse Tyson, director of the Hayden Planetarium, agreed, saying it looked like a new kind of musical instrument, an exotic flügelhorn, perhaps. Yes, I said, and it's a horn that plays itself!

In our model, there is no earliest event; every event has other events preceding it. Yet the Universe had a finite begin-

Figure 27. A Self-Creating Universe.
According to this model, in which universes give birth to other universes,
a time loop at the beginning allows the Universe to be its own mother.

ning. Specifically, in the time loop at the bottom, every event is
preceded by events lying counterclockwise from it in the loop.
Suppose we live in the universe on the far right, which we take
to represent a universe far out on the tree. Given an infinite
number of branches, we are likely to live in one that formed
much later than the first universe. Tracing back in time, we
would go down our branch into the universe to our left, then
into the trunk of the universe second from the left, then into
the loop at the bottom of the diagram, and then around and
around the loop forever. In the same way, the curved surface of
Earth has no easternmost point. You can continue to travel east
around Earth, and yet it is finite. If Earth were flat, as the
ancients believed, then it would either have an easternmost
edge or extend an infinite distance to the east. But because it is
curved, it can be finite and still have no easternmost point. Similarly, because general relativity allows curved geometries, we

can have a Universe that has *a beginning without having an earliest event.*

It caused itself.

Those who asserted that the Universe must either have a first cause or have existed infinitely back in the past did not envision curved spacetimes. This solves the first-cause problem in a way that would have been impossible to understand before general relativity.

Our model contains a Cauchy horizon separating the region of time travel from the later regions that have no time travel. This horizon circles the trunk just after the time loop branches off. If you live before that, you are in the time loop and can travel locally toward the future by going (clockwise) all the way around the time loop to return to your own past. But if you live after the horizon, you cannot visit your past. If you live to the future of the point where the time loop branches off, you just continue toward the future, going higher and higher on the tree. You can never get back to that loop at the bottom of the diagram. A time machine operates at the beginning of the Universe, but then it shuts down.

Our paper, "Can the Universe Create Itself?" was published in *Physical Review D* (the premier journal for particle physics) in May 1998. It has 155 equations and 187 references, yet its key idea can be summarized by Figure 27. Most of our paper is devoted to showing that we can find a self-consistent quantum vacuum state for the model, in agreement with Einstein's equations. We could find a self-consistent solution if the time loop had a particular length—one that, in fractions of a nanosecond, equals the initial circumference of the de Sitter space branch, in fractions of a foot. In this case, the negative energy density of our Rindler vacuum and the positive energy density from its being wrapped around a closed loop of time cancel each other exactly, leaving a pure inflationary vacuum state with a positive energy density and negative pressure everywhere—exactly what's required to produce the de Sitter geometry that we started with. This uniform vacuum state doesn't blow up on the Cauchy horizon or anywhere else. It is a self-consistent solution.

By the time the branch has circled around to become the trunk, its circumference has grown by a factor of $e^{2\pi} = 535.4916555\ldots$ To visualize this, imagine a tree trunk with a circumference of 535 inches, having a branch of 1-inch circumference sprouting from it. Then let the branch circle around and grow up to be the trunk. The length clockwise around the time loop is about 5×10^{-44} seconds. The density of the self-consistent vacuum state is about the Planck density—5×10^{93} grams/cm^3. This is just the density at which one calculates that quantum gravity effects should surely become important. We have no theory of quantum gravity at present, but it seems clear that, at such high densities and on such short time scales, quantum uncertainties in the geometry become critical. Spacetime is no longer smooth but becomes a complex, spongy tangle of loops called the *Planck foam*. This effect should make time loops like the one Li-Xin Li and I proposed even easier to make—in fact, almost hard to avoid. We also found that if a cosmological constant is associated with the unification of the strong, weak, and electromagnetic forces, then another self-consistent solution is possible, having a time loop about 10^{-36} seconds long. In this case, the density is well below the Planck density so quantum gravity effects should be unimportant, and our current calculations should be adequate as they stand. In either case, the time loop is extraordinarily short. Although we have no theory-of-everything yet, the general features of our calculation suggest that a small time machine at the beginning of the Universe is an attractive possibility.

Our solution also seems stable. This has been confirmed by calculations performed by Pedro F. González-Díaz, working at Hawking's home institute in Cambridge University. He found the solution stable against all fluctuations if the time loop is short—about 5×10^{-44} seconds.

After our paper appeared, we received many nice e-mails from our colleagues. John Barrow, one of the world's experts in the field called anthropic cosmology, wrote us that he had mentioned the possibility of closed timelike curves in the universe in a 1986 paper of his. Still, in that paper he called the time-travel alternative unappealing. Perhaps that's because before Kip Thorne's work, people did not take time-travel solutions that seriously.

The scenario Li-Xin Li and I have proposed is, like inflation, a general paradigm. Any scheme in which baby universes are produced can be changed into our type of model if one of those baby universes simply turns out to be the universe you started with. At MIT, physicists Edward Farhi, Guth, and Jemal Guven proposed that baby universes could be created in the lab by a supercivilization. According to this idea, one would compress a 10-kilogram sphere of mass to an extremely high density. This ball would go into a high-density vacuum state, whose negative pressure (or suction) would cause the ball to implode. While this would usually form just a black hole, occasionally it would branch off by quantum tunneling to create a baby universe hidden inside the black hole. This branch could grow up to a large size without interfering with the lab (the trunk universe). Edward R. Harrison of the University of Massachusetts has carried this notion further, proposing that our particular universe could have been created as a baby in a lab by some previous intelligent civilization. Harrison stated this might explain why the physical constants in our universe are conducive to intelligent life—they are simply similar to those in the parent universe, which harbored the intelligent civilization that created our universe. He suggested that all the baby universes, through many generations, could be formed this way, but he still needed a different, natural explanation to account for the first one, the trunk. With a time-travel loop, an intelligent civilization could produce the trunk as well.

Of course, this may overestimate the importance of intelligent civilizations. Natural formation of baby universes, as in Linde's chaotic inflation, may be far more common than creation by intelligent civilizations. Physicist Lee Smolin of Pennsylvania State University has suggested that every time a black hole forms, a baby universe is produced, branching off from our own and hidden from our view inside the black hole. If our universe possesses a tiny cosmological constant today, then Garriga and Vilenkin have shown that bubbles of high-density vacuum will eventually form. (Imagine a bowling ball on a coastal plain suddenly quantum-leaping into a mountain valley.) Each of these bubbles will branch off to create a separate inflating universe. This process is a version of Linde's chaotic inflation, in which random quantum fluctuations cause the formation of branching, inflating regions. Such a model could also be incorporated into our scheme.

Our idea that the Universe could create itself fits in very well with superstring theory, which proposes that, early on, all spatial dimensions were curled up and small. In our model's time loop, all dimensions—including time—are tightly wound and tiny. Our idea also meshes well with inflation. For the Universe to create itself through time travel, the Universe at some later time must resemble itself at some earlier time. Inflation allows this. If you start out with just a tiny bit of inflating vacuum, it will expand to enormous volume, little bits of which are exactly like the bit you started with. If one of these turns out to be the bit you started with, the Universe is indeed its own mother. Something remarkable happened at the beginning of the Universe—perhaps this was it.

THE ARROW OF TIME

Our model provides a resolution to an extraordinary paradox that has long captivated physicists—the arrow of time. Trade past for future, left for right, particles for antiparticles, and the laws of physics would work just as before—there is nothing magic about the future as opposed to the past. For example, the laws of electromagnetism make no distinction between the future and the past. But we know that light waves, which obey the laws of electromagnetism, travel only to the future. If I shake an electron now, light waves will go out at the speed of light, and 4 years from now they will reach the star Alpha Centauri, 4 light-years away. We call waves heading toward the future *retarded waves*. See Figure 28; on the left, the world line of an electron goes straight up, except for a kink in the electron world line when it is shaken. Retarded light waves are emitted at this point and proceed in a V pattern upward to the left and right at 45 degrees, at the speed of light toward the future. But we never see light waves going toward the past, although Maxwell's equations of electromagnetism would equally well permit this alternative solution, in which I shake an electron now, and light waves would go backward in time, intersecting Alpha Centauri 4 years ago. We call waves going toward the past *advanced waves*, like those emitted from the shaken electron on the right of Figure 28, producing an upside-down V in the spacetime diagram. But since we never see advanced waves, something must prevent them. The fact that we see light waves traveling exclusively toward the future from shaken charges explains the normal causality we observe in our universe today, where causes precede effects. I shake an electron now, and electromagnetic effects occur later, producing an arrow of time.

The same is true of gravity. Gravity waves, ripples in spacetime moving at the speed of light, also progress toward the future. Nobel laureates Russell Hulse and Joseph Taylor of Princeton have observed two neutron stars orbiting each other, slowly spiraling inward, ever closer, exactly as expected if they emitted gravity waves toward the future. If they emitted an equal quantity of gravity waves toward the past, the situation would

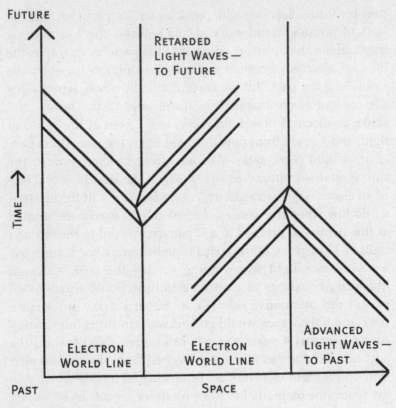

Figure 28. The Arrow of Time

be time-symmetric, and the stars would not spiral inward at all. If they emitted gravity waves only toward the past, it would then look like a movie of what we actually see, but played backward: we would see advanced waves converging on the binary stars (an upside-down V), giving them energy and causing them to spiral outward. But we see this pair spiraling inward, so gravity waves, as well as light waves, travel toward the future. This is puzzling.

In 1945, John Wheeler and Richard Feynman came up with an idea. They thought that electromagnetic waves from a shaken electron proceed in two directions: half to the past and half to the future. The waves going to the future eventually hit charges in the future, shaking them. These charges in turn would send out waves to the past and future. The waves these future charges send to the past ripple right back to the electron in the present, doubling the strength of the retarded waves emitted by the present electron, thereby bringing them to full strength. The waves from those charges in the future then continue on to the past, their crests and troughs exactly canceling the advanced waves that the electron had sent to the past—leaving no waves to the past of now. That would produce what we see. But why not the reverse—letting advanced waves shake charges in the past, so then the retarded waves of those past particles come back to cancel out the retarded waves—leaving just advanced waves? The time asymmetry we observe must ultimately derive from the existence of a very ordered (low-entropy) state with no waves, in the past, in the early Universe.

Do alternatives exist?

The geometry of our time-travel model provides a natural explanation for the asymmetry between the future and the past that we observe in our universe. Suppose we live in the universe represented by the farthest right horn in Figure 27. If we allowed light waves to go to the past, they would work their way back down this branch to the branch to its left and eventually to the trunk, where they would enter the time loop at the bottom and circle the time loop counterclockwise an infinite number of times, leading to an infinite buildup of energy and causing the whole structure to blow up, creating a singularity. That is not the geometry we started with—the solution is inconsistent. The only way for a self-consistent model to work is if light waves always travel toward the future, just as we observe. (If photons created in the branches travel only toward the future, then these photons travel out the branches away from the time loop, creating no problem.)

Now consider a photon emitted within the time loop at the bottom. It could, in principle, circle the loop clockwise an infinite number of times. But each time it went around, it would lose energy because it would be traveling toward the future, in the same direction that the branch is expanding. Each time it circled, it would add only one 535th as much energy as on the previous circuit because the expansion stretches its wavelength by a factor of 535, robbing it of energy. The sum rapidly converges to a finite value. So, even though it circles an infinite number of times, it would not cause an infinite buildup of energy. However, a photon going backward in time (counterclockwise) around the loop would pick up energy on each circuit because in the counterclockwise direction the branch is always getting smaller, compressing its wavelength. A photon circling an infinite number of times toward the past would cause an infinite buildup of energy, causing the model to blow up. In fact, the only way to obtain a self-consistent solution is to have light waves, and gravitational waves, travel only to the future throughout the entire model. Thus, in our model, the asymmetry between the future and the past that we observe (in which causes precede effects) comes from the time asymmetry in the geometry of Universe—it has a time loop in the beginning.

This arrow of time was not something we built into the model; it was implicit in the model, but its emergence quite surprised us. It's an important prediction by the model, which turns out to agree with our observations.

In the standard big bang model, by contrast, there is nothing to produce an arrow of time. In that model, the early universe is filled with radiation, and whether it is going forward or backward in time from its source does not matter. Waves going toward the past would increase in energy as they approached the big bang singularity, where they would blow up. But the density in the big bang model blows up there anyway, so it causes no problem. Waves going to the past are not forbidden, in principle, in the standard big bang model. But with a time loop in the beginning, self-consistency forbids waves going to the past—just as is observed today.

What about the "entropy arrow of time," the increase in disorder over time that we observe in the universe? It occurs because many disordered states exist, but only a few ordered states do. Here's an example of this principle in action. Place 100 coins carefully in a shoebox, all heads up. That is a highly ordered state—it took energy to check each coin. Shake the box. When you look in later, you will likely find some coins heads up and some heads down—a random or disordered state. There are many ways to have some coins heads up and some tails up—different coins can be heads up in each particular case—but only one way to arrange the coins so that all are heads up. There is always a small chance that you will look in after shaking and find all 100 coins heads up—1 chance in 2 raised to the hundredth power. Shake the box once per second; it would take 40 billion trillion years to likely turn all 100 coins heads up by chance.

Likewise, the probability of randomly encountering an ice cube on a hot stove, instead of water or steam, is so tiny that we would normally not expect to find it by chance. If we see an ice cube on a stove, it's typically because someone has placed it there. Then if we look 5 minutes later, we will find it half melted—a more disordered state. Relying on an argument presented by Wheeler and Feynman in their 1945 paper, let's see what would happen if that ice cube were not placed there by anyone, but just turned up as a very improbable statistical fluctuation—like randomly finding all the coins heads up. If we looked 5 minutes later, we would expect to find the cube half melted. But suppose we looked 5 minutes earlier. We should again expect to find the cube half melted—for finding a still larger ice cube at the earlier time would be even more incredibly unlikely than finding the first one.

Thus, the laws of physics do not establish an arrow of time based on entropy; they simply say that entropy increases as one moves away from a state of order (whether toward the future or the past). However, if the Universe finds itself already in an incredibly ordered state at the very beginning, we should expect to find it more and more disordered at later and later times.

In our time loop model, we can calculate the temperature at any point. We find that the entire volume within the time loop itself is cold—at a temperature of absolute zero. This is a highly ordered, low-entropy state. The loop of time is filled with a pure inflationary vacuum state of zero temperature. We find no particles there and no radiation. On the other hand, after the Cauchy horizon is passed, in the branches after the time loop, we find that the Universe is hot. (An observer in the region after the time loop would be immersed in a hot bath of Hawking radiation, since there are event horizons in the space. Event horizons are produced because the rapidly inflating nature of the space guarantees that light from very distant events will never reach the observer.) Going from cold to hot represents an increase in disorder. Thus, there is an *entropy arrow of time* (more disorder at later times) in our model that parallels the *electromagnetic arrow of time*. Because the Universe starts off automatically in a low-entropy state in our time loop, disorder should naturally spread from there, explaining why disorder increases with time today.

Our model therefore offers a new and promising idea for how the Universe began. It takes advantage of a remarkable property of general relativity to address the question of first cause in a novel way. In fact, time travel seems perfectly suited to resolving this problem.

Someone might ask about the theological implications of our model. Li-Xin Li and I have never discussed theology. We have no theological axes to grind. We simply wanted to see whether an interesting property of general relativity could be useful in explaining the origin of the Universe. That's a proper task for physicists. I would be reluctant to draw theological conclusions from our model; the results speak for themselves. A professional theologian might well point out that a self-creating Universe was interesting, but it still didn't answer the question of why there was a self-creating Universe rather than none at all. What people make of our self-creating model may depend on their outlook to begin with. Atheists and pantheists might well find a self-creating Universe attractive. As a religious person, I would not pretend that a self-creating Universe is not a troubling notion—but perhaps we should find the Universe troubling.

For a visual meditation on this theme, consider the Escher drawing in Figure 29.

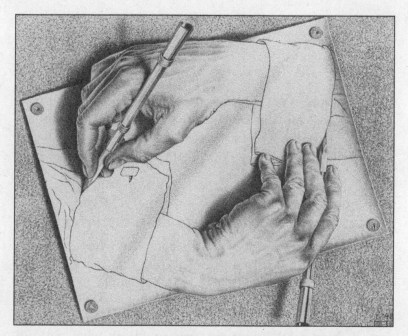

Figure 29. *Drawing Hands* (1948), by M. C. Escher.

5 REPORT FROM

THE FUTURE

Hope is not the conviction that something will turn out
well but the certainty that something makes sense,
regardless of how it turns out.

— VÁCLAV HAVEL, *DISTURBING THE PEACE*

QUESTIONS FOR A TIME TRAVELER FROM THE FUTURE

No book on time travel would be complete without a report
from the future. If a time traveler from the future suddenly
appeared, what would you ask her? You might like to know
how your current relationship will turn out, how the company
you work for will prosper, how long the country you live in will
last. Perhaps most important, you could ask what will ulti-

mately become of the human race. In fact, it was this very piece of information that H. G. Wells's Time Traveler brought back to his friends.

Could a warning from the future save us from some awful fate? It might, according to the many-worlds picture of quantum mechanics of David Deutsch described in Chapter 1. In that view, many possible futures exist, and a time traveler can simply return from one of them. If most future universes contained some catastrophe, then most time travelers from the future would report it. (A time traveler could then tell you only what was likely to happen.) If you heeded the warning, you might be able to avoid the catastrophe by moving into a future universe without it. Alternatively, if Thorne and Novikov are right, then a time traveler would report events that must surely happen in this world's future. Any warning such a time traveler might deliver could not, by definition, change the course of events. As Brandon Carter reminded me recently, this was the plight of Cassandra in ancient Greek mythology; she was given the ability to accurately forecast the future, but it came with the curse that her prophecies would not be believed. Can we get an accurate scientific prophecy of the future? Perhaps—if, once again, we ask the right question.

SCIENTIFIC PREDICTION OF THE FUTURE

Science has been in the business of making predictions about the future for a long time. Ancient Egyptian astronomers could predict the flood season of the Nile by noting when the bright star Sirius rose. Astronomers noted cyclic, repeating patterns in the sky and predicted that these patterns would continue into the future—and such predictions turned out to be correct. With more observations and sophistication, astronomers could predict future eclipses of the Sun. The Greek astronomer and mathematician Thales became famous for correctly predicting the solar eclipse of May 28, 585 B.C.E.

Using Newton's theory of gravity, one could predict the future motions of astronomical bodies from their current positions and velocities. In 1705, Edmund Halley used Newton's theory to determine that the comet he observed in 1682 would next return to Earth's vicinity in about 1758. Halley died before that date, in 1742, at the age of 85. But when the comet did return as he had predicted, people named it after him.

Newton's theory of gravity formed the basis of thousands of successful predictions. But when it failed to explain the precession of the orbit of Mercury and the bending of light around the Sun, it was supplanted by Einstein's more accurate theory of gravity. The scientific method works because it doesn't shrink from replacing even a great theory like Newton's if it makes incorrect predictions.

As science progressed, it became ever more accurate at predicting future events. When comet Shoemaker-Levy was discovered in March of 1993, astronomers used its position and velocity to correctly predict that it would collide with Jupiter a little more than a year later. This enabled astronomers to be ready to observe the event through ground-based telescopes around the world and with the Hubble Space Telescope. Similarly, the physics of hydrodynamics enables meteorologists to forecast the weather accurately several days ahead, giving advance warning of hurricanes and blizzards and thereby saving lives. These are all scientific predictions using well-defined methods whose success has been checked in the past.

Indeed, during the time of Newton and after, scientists hoped that their ability to predict the future would improve without bound. According to Newton's theory, if one knew the mass, position, and velocity of every particle in the universe, one could calculate each one's positions as far into the future as one wanted. Thus, if one could obtain accurate knowledge about just the present, one would be able to predict all of the future as well. This was the vision of a clockwork universe.

But the Heisenberg uncertainty principle in quantum mechanics says that we cannot simultaneously measure both the position and velocity of any particle with arbitrary accuracy —we cannot achieve the Newtonian dream of knowing exactly the position and velocity of every particle in the universe at the present epoch. Therefore, a detailed, perfect prediction of the future of all these particles is impossible in principle.

Even worse, chaos theory tells us that many dynamical systems are chaotically unstable. This means that small uncertainties in the positions and velocities of particles will propagate into the future, becoming larger and larger until our predictions no longer resemble the actual course of events at all. We can forecast orbits of near-Earth asteroids accurately for only about a hundred years before chaos sets in and our solutions become meaningless. Many important systems are chaotic. Weather is chaotically unstable over a period of a few days. This is epitomized by the famous statement that a butterfly flapping its wings in the Amazon Basin can change the course of a hurricane months later in the Caribbean. Small changes accumulate into larger changes, doubling again and again. To properly calculate the weather many months ahead would require an impossibly accurate picture of the current weather and the ability to forecast the movement of every animal on Earth. Thus, we have at best a short horizon for making detailed weather predictions.

Biological evolution also appears chaotic. Go back in time, kill one extra trilobite 500 million-million years ago, and perhaps human beings would never evolve—evolution could simply spin off in a different direction. Five hundred million years ago, one could never have predicted how a Tyrannosaurus would look or the physical appearance of human beings. Harvard's Stephen Jay Gould has eloquently argued this case in his book

Wonderful Life. Rewind the tape of history and replay it, and it could play out completely differently in detail.

In many of the matters we care most about, including the fate of our own species, our ability to make detailed Newtonian-type forecasts appears hopeless. This has caused many people to pronounce that the future is absolutely unpredictable. This is unduly pessimistic. Prior to the arrival of any time travelers from the future, what can we say about it? Quantum mechanics tells us that in principle all predictions we make about the future must be stated in terms of probabilities of outcomes of future observations. But in fact, these probability estimates can be extremely useful and may tell us all we can really know.

Knowing that the future of the universe is not calculable in detail, therefore, does not mean that we cannot make predictions about it. For example, I can predict that it will snow sometime next year in New York City and be pretty confident of being correct. This is a different type of prediction, a statistical prediction, that does not require me to follow each weather front in detail and is not disrupted by that butterfly near the Amazon.

Often scientists are called upon to predict the future, not by applying a particular scientific hypothesis, as Halley did using Newton's theory, but simply as experts, knowledgeable in science and the laws of physics, who are asked to prognosticate about the future. Often this involves knowing that X would not violate the laws of physics and then predicting that our technology will advance to allow us to do X. In the 1890s, when Russian physicist Konstantin Tsiolkovsky predicted that people would go into space with rockets, his forecast fell into this category. Likewise, Jules Verne predicted correctly the first nuclear submarine. But Verne also predicted that people would journey far into Earth and find dinosaurs living there—which hasn't happened yet. While such prognostications can occasionally turn out to be spectacularly correct, they often remain unfulfilled. In 1974 Gerard O'Neill of Princeton forecast that the number of people living in space by 1996 could be in the range of 100,000 to 200,000. He even proposed how to accomplish this feat—by building large space colonies. The idea was good, but it wasn't carried out.

The trouble with such prognostications is that they are just educated guesses. History has shown that they can be wildly wrong and often wildly optimistic, particularly in the case of risk assessment. Nuclear power plants were supposed to be so safe that the chance of an accident approximated that of your being hit by lightning—then Three Mile Island and Chernobyl proved that surmise wrong. What causes a failure in the end is usually something surprising, something unforeseen in the calculations, making the ultimate failure rate higher than we had supposed.

When Mrs. Albert Caldwell boarded the *Titanic*, she asked a deckhand, "Is this ship really nonsinkable?" "Yes, lady," he replied. "God himself could not sink this ship." This prognostication, recorded in Walter Lord's *A Night to Remember*, was based on the fact that the *Titanic* was a new ship, built with 16 watertight compartments. If a leak occurred, that compartment could be sealed shut and the ship would not sink. This safeguard seemed pretty foolproof. Of course, the unforeseen did happen. A spur on the iceberg that the *Titanic* struck scraped along the side of the ship under water, popping open plates along a 300-foot length. Similarly, the supposedly invincible German battleship *Bismarck* went down on its maiden voyage as well. Precisely because the *Bismarck* was deemed invincible, the British feared it and sent almost their entire fleet after it, and sunk it. Prognostications can often be wrong.

I am now going to make some predictions. They are not prognostications—not the opinions of just one expert whose hopes or fears can be argued and weighed against those of other experts, as in most futurist books. Instead they are, in the tradition of scientific predictions such as Halley's, based on a particular scientific hypothesis, one that has been astonishingly successful in the past. They will tell you how long the human race is likely to last and how you could have known, at the time, to have stayed off the *Titanic* and the *Bismarck*.

PREDICTING THE FALL OF THE BERLIN WALL

In 1969, while standing at the Berlin Wall, I discovered a way of predicting how long something you are observing is likely to last. This is based on the *Copernican principle*, the idea that your location is not special, one of the most famous and successful scientific hypotheses of all time. It's named after Nicolaus Copernicus, who proved to people that Earth did not occupy a special location at the center of the universe. Our subsequent discoveries—that we orbit an ordinary star in an ordinary galaxy in an ordinary supercluster—continue to make our position look ever less special. The Copernican principle works because, by definition, out of all the places for intelligent observers to live, only a few special places and many more nonspecial places exist. You are simply likely to be at one of the many nonspecial places. Christiaan Huygens (Newton's clever contemporary, who developed the wave theory of light and the most accurate clock of his day) used this principle to correctly predict the distances to the stars. He reasoned, Why should the Sun be special, the brightest light in the universe? He noted that if Sirius, the brightest star seen in the sky, was intrinsically as bright as the Sun, he could figure out its distance simply by estimating how far away you would have to move the Sun to make it look as dim as Sirius. Later investigators found that Huygens had gotten the distance to Sirius right to within a factor of 20, a remarkable accomplishment for that day.

When Hubble discovered a distribution of galaxies that was equal in all directions and expanding away from us, we could have interpreted this as resulting from our being located at the center of the universe. If we are not special, however, it must look that way to everyone, thus leading us to the standard big bang models that Gamow, Herman, and Alpher used to predict the existence of the cosmic microwave background radiation. This remains one of the most remarkable predictions to be verified in the history of science—all because of taking seriously the idea that your location is not special.

In 1969, at the time of my visit to the Berlin Wall, it had been standing for 8 years. People wanted to know how long the wall was going to last. Some people thought it would be a temporary aberration, while others thought it would remain a permanent fixture of modern Europe. I reasoned, using the Copernican principle, that since there was nothing special about my visit, I was simply observing it at some random point during its existence—somewhere between its beginning and end. If there was nothing special about the location of my visit in time, there was a 50 percent chance that I was observing the wall sometime during the middle two quarters of its existence. If I was at the beginning of this middle interval, then one quarter of the wall's existence had passed and three quarters remained in the future. If I was at the end of the middle two quarters, then three quarters of its existence had passed and only one quarter lay in the future. Thus, there was a 50 percent chance that the future longevity of the wall was between $\frac{1}{3}$ and 3 times as long as its past longevity (see Figure 30). Now 8 years divided by 3 is $2\frac{2}{3}$ years, while 8 years multiplied by 3 is 24 years. So standing at the wall in 1969, I predicted to a friend, Charles Allen (now president of the Astronomical League), that there was a 50 percent chance that the future longevity of the wall would be between $2\frac{2}{3}$ years and 24 years. I made no prediction of why the Berlin Wall would end, just how much longer it was likely to last. My prediction could easily have been incorrect. The Berlin Wall could have been destroyed by a nuclear

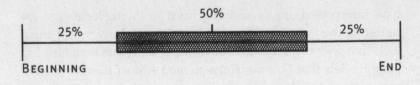

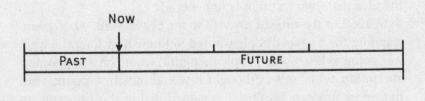

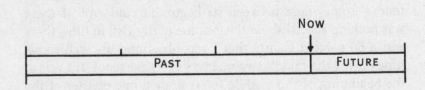

If you observe something at a random time, there is a 50 percent chance you will catch it in the middle two quarters of its period of observability (top diagram). At one extreme (middle diagram), the future is 3 times as long as the past, whereas at the other extreme (bottom diagram), the future is one third as long as the past. There is a 50 percent chance that you lie between these two extremes and that the future is between one third and 3 times as long as the past.

Figure 30. The 50 Percent Copernican Argument

weapon milliseconds after my prediction had been made (this was during the Cold War, after all), or it could have lasted for thousands of years.

But 20 years later, I called my friend. I said, "Chuck, you remember that prediction I made about the future longevity of the Berlin Wall?" He did. "Well, turn on your TV because Tom Brokaw is at the wall now, and they are tearing it down!" When the wall came down in 1989, after 20 years, in agreement with my original prediction, I decided that I should write this up.

The Future of the Human Race

I wanted to apply this technique of using the Copernican principle to something important—predicting the likely future longevity of the human race. That became the main thrust of "Implications of the Copernican Principle for Our Future Prospects," my paper that appeared in *Nature* on May 27, 1993. Of course, scientists like to make predictions having more than a 50 percent chance of being correct. Scientists often adopt the criterion of making predictions only when they have at least a 95 percent chance of being correct—high enough that betting against the prediction is like backing a long shot at the track, but low enough to set interesting limits. It has become the standard for scientific predictions.

How does this change my argument? Assuming nothing is special about your location, when you observe something, there is a 95 percent chance that you are seeing it during the middle 95 percent of its period of observability (that is, you are not in either the first 2.5 percent or the last 2.5 percent of the time interval when it can be seen). The end comes either when whatever you are observing is destroyed or when there are no longer any observers left to observe it, whichever comes first. Now 2.5 percent is $\frac{1}{40}$th. If you are at the earliest point of the middle 95 percent, you are just 2.5 percent from the beginning; in that case, 1/40th of the interval is in the past and 39/40ths of the interval lie in the future—the future is then 39 times as long as the past. At the other extreme, if you are at the end of the middle 95 percent, you are situated just 2.5 percent from the end. In that case, 39/40ths of the interval are in the past and 1/40th of the interval remains in the future, making the future only 1/39th as long as the past. Thus, you can say with 95 percent confidence that you fall between these two extremes and that the future longevity of whatever you are observing lies between 1/39th and 39 times its past longevity (see Figure 31).

Our species, *Homo sapiens*, has been around for about 200,000 years. If there is nothing special about our time of observation now, we have a 95 percent chance of living sometime in the middle 95 percent of human history. Thus, we can set 95 percent confidence level limits on the future longevity of our species. It should be more than 5,100 years but less than 7.8 million years (5,100 years is 1/39th of 200,000 years and 7.8 million years is 39 times 200,000 years). Interestingly, this gives us a predicted total longevity (past plus future) of between 0.205 million and 8 million years, which is quite similar to that for other hominids (*Homo erectus*, our direct ancestor, lasted 1.6 million years, and *Homo neanderthalensis* lasted 0.3 million years) and mammal species generally (whose mean longevity is 2 million years). The average, or mean, duration of all species lies between 1 million and 11 million years.

Some might claim that as an intelligent species—one able to reason abstractly, create art, ask questions such as "How long will our species last?" and so on—the normal rules of extinction do not apply to us. In theory, we could use our discoveries to better our position through genetic engineering (to alter ourselves as needed) or through space travel (to vastly expand our habitat). High technologies, however, also pose substantial risks,

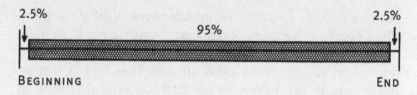

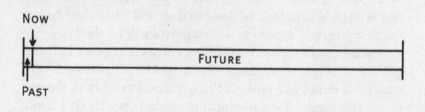

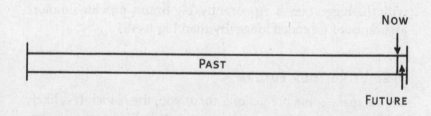

If you observe something at a random time, there is a 95 percent chance you will catch it in the middle 95 percent of its period of observability (top diagram). At one extreme (middle diagram), the future is 39 times as long as the past, whereas at the other extreme (bottom diagram), the future is $\frac{1}{39}$th as long as the past. There is a 95 percent chance that you lie between these two extremes and that the future is between $\frac{1}{39}$th and 39 times as long as the past.

Figure 31. The 95 Percent Copernican Argument

such as biological warfare or missile-borne nuclear weapons. The Copernican estimate of our future longevity is based only on our own past longevity as an intelligent species and does not depend on data from any other species. Therefore, it is noteworthy that its predictions for our total longevity are similar to the longevities observed for other species. If we remain on Earth, we will also be exposed to many of the same risks that other species face, including major epidemics, climatological and ecological disasters, asteroid strikes, and so on, and so one might argue that, therefore, our longevity will be similar.

Unfortunately for us, no positive correlation exists between general intelligence and longevity of species. Einstein was very smart but didn't live orders of magnitude longer than the rest of us. The species *Tyrannosaurus rex* lasted only about 2.5 million years. It was the most fearsome predator up to its time, with the biggest teeth. Apparently, big brains provide no more assurance of extended longevity than big teeth.

PREDICTING YOUR FUTURE

Let me make some predictions about you, the reader. It is likely that you were *not* born on January 1. It is likely that you are located in the middle 95 percent of your hometown phone book—in the United States this means somewhere between Aona and Wilson. You are likely to have been born in a country whose population is larger than 5.8 million. Are most of these predictions, perhaps all of them, right? I surmise these things simply because I assume there was nothing special about your location at birth.

Any good scientific hypothesis should be testable, and the Copernican principle is no exception. Fortunately, it provides numerous predictions that can be tested—many in everyday life. The day my *Nature* paper came out, May 27, 1993, I checked *The New Yorker* to find all the Broadway and off-Broadway plays and musicals open at that time—and found 44. Calling up each theater, I found out how long each play or musical had been open. Then I waited to see how long each took to close. I chose to test plays for two major reasons. First, since most of the plays had not been open long, I was likely to get interesting results before many years had passed. Second, the durations of Broadway plays are notoriously difficult to predict. The star may die. The theater may burn down. Or a new star may extend the run. Plays are subject to chaotic uncertainties, just like species are. Therefore, they constitute an important test.

Thirty-seven of these plays and musicals have now closed— all in agreement with the 95 percent confidence-level predictions of my formula. For instance, the *Will Rogers Follies*, which had been open for 757 days, closed after another 101 days, and the *Kiss of the Spider Woman*, open for 24 days, closed after another 765 days. In each case, the future longevity was within a factor of 39 of the past longevity, as predicted.

How long did people think these runs would last—or at the least, how long did the shows' publicity engines suggest they would last? Newspaper advertisements for *Kiss of the Spider Woman* promised, "The Kiss that lasts forever." But shortly before it closed, new ads appeared offering just "One Last Kiss"! Ads for *Cats* at the time boasted "*Cats*—Now and Forever." When my paper came out, *Cats* had been open for 10.6 years. It closed 7.3 years later. How did I know *Cats* would not last forever? If it lasted forever, then all but an infinitesimal fraction of its observers would find it nearly as old as the universe itself, but my noticing that it is many orders of magnitude younger than the universe would make the timing of my observation very special.

The same reasoning tells us that *Homo sapiens* (presumably the first *intelligent* species on Earth—that is, able to ask questions like this) and its intelligent descendants, if any, will not last forever either. We observe that our intelligent lineage is

200,000 years old in a universe that is 13 billion years old. The ratio of these two ages is 1:65,000. But suppose human beings and their intelligent descendants were to last forever. People a trillion years from now would observe an age for their intelligent lineage of 1,000,000,200,000 years and an age for the universe of 1,013,000,000,000 years. The ratio between these two ages is 0.987, a number near 1. The infinite number of people living after that would see a ratio even closer to 1. Thus, if humans and their intelligent descendants lasted forever, all but an infinitesimal fraction of them would observe that ratio to be a number near 1. But you would observe 1/65,000, a number much less than 1, so that would make you very special (see Figure 32). Thus, neither the human race nor its intelligent descendants are likely to last forever. Being assured that they do have an end, we can show with 95 percent confidence where that end lies: between 5,100 years and 7.8 million years in the future. In all these applications, we are ultimately relying on the Copernican idea that your observation is not likely to be special among similar observations.

On September 30, 1993, in *Nature*, P. T. Landsberg, J. N. Dewynne, and C. P. Please used my formula to predict how long the Conservative government in Britain would continue in power. Since the Conservative Party had been in power for 14 years in 1993, they estimated with 95 percent confidence that it would remain in power for at least 4.3 more months but less than 546 more years. The Conservative Party went out of power 3.6 years later, on May 2, 1997, in agreement with the prediction.

After my paper had been published I received a nice note from Henry Bienen, then dean of Princeton's Woodrow Wilson School. He noted that he and Nicholas van de Walle had written a book in 1991, *Of Time and Power*, which, after a detailed statistical study of 2,256 world leaders, concluded, "The length of time that a leader has been in power is a very good predictor of how long that leader will continue in power. Indeed, of

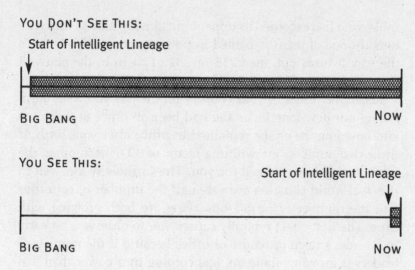

YOU DON'T SEE THIS:

Start of Intelligent Lineage

BIG BANG   NOW

YOU SEE THIS:

Start of Intelligent Lineage

BIG BANG   NOW

If humans and their intelligent descendants were to last forever, then all but an infinitesimal fraction of them would observe (top diagram) their intelligent lineage (shaded) to be nearly as old as the universe itself (back to the big bang). But you observe (bottom diagram) that our intelligent lineage is much younger than the universe. That would make you very special, and that's not likely.

Figure 32. Why We Are Not Likely to Last Forever

all the variables examined, it is the predictor that gives the most confidence."

Of the 115 world leaders in power at the time of my birth (February 8, 1947), 108, or 94 percent, had their future longevities in office predicted correctly by the 95 percent Copernican formula—a pretty good outcome. My student Lauren Herold replicated this experiment. Of the 232 world leaders in power on the date of her birth (March 12, 1975), 209 leaders had gone out of power by the time she completed her survey in 1996. The 95 percent Copernican formula correctly predicted 196 of these. The 23 leaders still in power included 1 sure loser for the formula, who had exceeded its upper limit already, and 22 sure winners if none of them remained in power past age 150. If that's the way it turns out, then 218 of 232 of them in the end will have been correctly predicted, for a success rate of 94 percent.

Recall the leader of your country on the day you were born. Figure out how long he or she had been in office at that time and how long he or she remained in office after your birth. If these two numbers are within a factor of 39 of each other, the formula will have worked for you. The formula works well in this real-world situation even though the number of countries and the number of people observing are both growing with time. The first effect typically causes you to observe a bit early in a leader's reign or tenure of office because if the number of leaders is growing, more are just coming into power than just going out; the second effect typically causes you to observe a bit late because more people are alive to observe the latter part of any leader's reign or tenure of office. The two effects tend to cancel each other out. They will do so as long as the number of items being observed is proportional to the number of people observing, as might seem reasonable, since each person can observe only so much.

After I gave a talk on this topic at the annual meeting of the Astronomical Society of the Pacific, someone asked me what my paper had predicted for my own future longevity. The answer: At the time my paper was published on May 27, 1993, I was 46.3 years old, so the 95 percent formula predicted that I should live at least another 1.2 years but less than another 1,806 years. I've survived past the lower limit, so assuming I don't make it to the upper limit, the formula will have worked for me. Of all the people alive when my paper was published, the 95 percent Copernican formula should predict their future longevity correctly in 96 percent of the cases, applying the 1983 worldwide actuarial and population distribution tables of Ansley Coale and his colleagues, appropriate for the life expectancy, rate of population growth, and population distribution with age found in 1993. Using those actuarial tables, one can predict the fraction of people in each age group for whom the formula will work. This fraction is greater than 95 percent for youths through middle-aged adults, and lower than 95 percent for babies and very old people. Since you are unlikely to be among the very youngest or oldest people alive today, the formula is likely to work for you.

Of course, you could get a narrower range for your future longevity by just using actuarial tables, taking advantage of the fact that you know not only your own age but also the ages at which billions of other people have died. With the help of this larger database, and again applying the Copernican principle, you can assume that you are not special among human beings and obtain an improved estimate. But if you lived on a desert island and had never known of any other human beings, the 95 percent Copernican formula would have allowed you, using only your current age, to make a rough estimate of your future longevity, with 95 percent confidence of accuracy. Since we have no actuarial data on intelligent species other than our own, the 95 percent Copernican estimate for the future longevity of our species is arguably the best we can make.

Now for some historical applications.

When I visited the Soviet Union in 1977 and walked around Red Square, I remember thinking to myself that since the Soviet Union was only 60 years old at the time, it might not last as long as many people thought it would. The major threats to its existence prior to my visit, including attack by Nazi Germany and the threat of nuclear war during the Cold War, were either gone or lessening, and many argued that its future stability in a sort of infinite standoff with the United States seemed assured. But 14 years later it was gone. I presume that my visit did not bring about the fall of the Soviet Union and, at the time of my visit, predicting the exact causes of its future demise would have been impossible. Glasnost and perestroika could not have been anticipated in 1977—they were not even in our Cold War vocabulary. The 95 percent Copernican argument worked—the future was within a factor of 39 of the past —even though the rules were changing and the threats in the future would be unlike those seen in the past. This is simply because, in the end, my visit did not turn out to be special.

In 1956 Nikita Khrushchev boasted, "We will bury you." This was taken as a rather ominous warning that he planned to destroy the United States, but actually the comment came from an old Russian proverb meaning simply that we will outlast you, we will attend your funeral, we will be there to bury you. The assertion was boastful because at that time the Soviet Union was only 39 years old whereas the United States was then 180 years old. Indeed, the Soviet Union was gone in another 35 years and the United States outlasted it.

It is dangerous to make predictions that fall outside the factor-of-39 limits implied by the Copernican argument. In 1934, after being in power only 1 year, Adolf Hitler made a very famous, and ominous, prediction that there would be no further revolutions in Germany in the next thousand years. His prediction that the Third Reich would last for a thousand more years scared people around the world. Fortunately, the prediction was rash, for the Copernican argument would have predicted with 95 percent confidence that the future longevity of the Third Reich would be more than 9 additional days but less than 39 more years. In agreement with this prediction, 11 years later both Hitler and the Third Reich were dead.

The famous list of the Seven Wonders of the World can be traced back to approximately 150 B.C.E., the time of Antipater of Sidon. Two of the Seven Wonders (the Hanging Gardens of Babylon and the Colossus of Rhodes) no longer existed at the time the list was made, but five still did: the statue of Zeus at Olympia, the temple of Artemis at Ephesus, the mausoleum at

Halicarnassus, the Pharos of Alexandria, and the pyramids of Egypt. Of the first four wonders that had each been in existence for less than 400 years at the time the list was made, not one is still here today. But the oldest, the pyramids, which were then 2,400 years old, have survived. Things that have been around for a long time tend to stay around a long time. Things that haven't been around long may be gone soon.

Write down the precise time and date that you are reading this sentence: _____ (year) _____ (month) _____ (day) _____ (hour) _____ (minute) _____ (second). The publication of my book has nothing special to do with you. So the time above, at which you read the first sentence of this paragraph, should be at a nonspecial time with respect to matters important to you. Using the above time as your observation point, you can use the 95 percent Copernican formula right now to forecast the future longevity of your current relationship, the country in which you live, the college you attended or hope to attend, the company you work for, or your favorite magazine.

When should you *not* use the formula? Don't wait until you are invited to a friend's wedding, and then, 1 minute after the vows are finished, proclaim that the marriage has less than 39 more minutes to go. You attended the wedding precisely to observe a *special* point in the marriage—its beginning. But you can use the time you filled in above as the observation point to predict the future of your own marriage (if you are already married) because your reading that sentence is unrelated to your marriage and is likely to occur at some random point within it.

I would not use the formula to predict the future longevity of patients in a nursing home since a nursing home, by definition, assists people at a special time near the end of their lives. Using the past duration of a patient's stay in the home, however, you could apply the formula to predict the length of the rest of their stay.

Do not use my formula to predict its own future longevity.

My paper and papers written by people who were present in 1993, like guests at a wedding, are located by definition at a special place in the history of when my formula will be known— near its beginning. My paper may cease to be known in the future, not because it is ever shown to be wrong, but simply because it is forgotten. Aristarchus correctly argued in 260 B.C.E. that Earth revolved around the Sun, but his book was lost and his work was largely forgotten until Copernicus came along.

Do not apply the formula to predict the longevity of the universe. Intelligent observers were not present at its beginning. Since they may also die out long before the universe, your observation point may be special with respect to universal history. Intelligent observers live in a habitable (therefore, perhaps special) epoch in that history (an idea called the "weak anthropic principle"). Your viewpoint, however, should not be special *among* intelligent observers. For things older than the human race, mark the beginning with the first human observations of them and the end with the last human observation of them. We are just predicting the period of future observability from the period of past observability by relying, as in all these applications, on the hypothesis that your observation is not special among similar observations.

Quantum mechanics tells us that observing a system can influence it. If you make a prediction about something unimportant and easy to change (for example, how long you have been wearing the clothes you have on now), you could make the prediction wrong simply by taking off all your clothes right now. If you are standing in a bookstore, this could be embarrassing, but if you are reading at home, you could do it easily. For something important, such as your marriage, you would not end it immediately just to prove a prediction wrong. Thus, this effect is unlikely to be significant for matters of importance. For example, the *Nature* article predicting the fall of the Conservative government in England could in principle have *caused* the Conservative MPs to make a no-confidence vote that day just to prove the prediction wrong—but that's not likely, and it didn't happen. They wanted to continue in power as long as they could, regardless of predictions made about them.

Rachel Silverman, a reporter with the *Wall Street Journal*, called me in the fall of 1999 and asked me to make a series of predictions for their issue of January 1, 2000, which would focus on the future. That is a very special day as far as the calendar is concerned, but it may well be nonspecial relative to our observation of other things. Here are the matters she chose (her decisions, not mine) for their interest to her and *Wall Street Journal* readers, along with their 95 percent confidence level predictions, which were published on January 1, 2000:

PHENOMENON AND ITS STARTING DATE	FUTURE LONGEVITY	
	More Than	But Less Than
Stonehenge (2000 B.C.E.)	102.5 years	156,000 years
Pantheon (126 C.E.)	48 years	73,086 years
Humans (*Homo sapiens*) (200,000 years old)	5,100 years	7.8 million years
Great Wall (of China) (210 B.C.E.)	56 years	86,150 years
Internet (1969)	9 months	1,209 years
Microsoft (1975)	7 months	975 years
General Motors (1908)	2.3 years	3,588 years
Christianity (c. 33 C.E.)	50 years	76,713 years
United States (1776)	5.7 years	8,736 years
New York Stock Exchange (1792)	5.2 years	8,112 years
Manhattan (purchased in 1626)	9.5 years	14,586 years
Wall Street Journal (1889)	2.8 years	4,329 years
New York Times (1851)	3.8 years	5,811 years
Oxford University (1249)	19 years	29,289 years

By "Manhattan" one means New York City, because it was the city that started there in 1626. The Internet could end either because that technology ends or because it is replaced by something better (*Star Trek*'s Holodeck, for example). As usual, the end for Stonehenge, the Pantheon, the Great Wall, and so forth would occur either when they are torn down or disappear by other means or when there is no one left to observe them.

Make your own list. If you use the formula to predict 100 things about your future, picked at random, remember that on average about 5 of them should turn out wrong. Choose a half dozen things that are most important to you, and all those predictions might turn out right.

The argument can be useful in everyday situations, particularly when you are traveling. To be on the conservative side, if you go to the dock to take an ocean voyage, don't pick a ship that has not already completed at least 39 such voyages successfully. This will keep you off particularly unfortunate ships. This simple rule would have kept you off the *Titanic* and the *Bismarck* (it leaves you off maiden voyages; interestingly, the Vanderbilts canceled passage on the *Titanic* because Vanderbilt's mother had a personal rule against maiden voyages). It would also have kept you off the *Hindenberg*, which exploded on its 35th transatlantic voyage, and the *Challenger* space shuttle, which met disaster on its tenth flight. A long, successful track record is a good safety advertisement, proving the vessel has survived all possible catastrophes for a large number of outings. If you arrive at a random time at the dock and find a ship with a long series of successful voyages behind it, the Copernican principle would suggest that its next voyage is not likely to be its last.

When I was in Hong Kong, I wanted to ride the inclined railway to the top of Victoria Peak. It looked pretty steep, so I asked the ticket taker how long it had been since the last accident. He said the railway hadn't had an accident in the 90 years since it opened. I got on.

MAY YOU LIVE IN INTERESTING TIMES

Any species starts with just a few members, reaches a maximum population at some point, and then typically declines to just a few individuals before going extinct. Where should you expect to find yourself in such a population curve? Near the peak, of course—because most individuals live then, so they (and you) are not special. The Copernican principle indicates that you are likely to be born in a century with a population higher than that of the median century. Why? For the same reason you are likely to be born in a country with a population larger than that of the median country—because most people are. Half the world's 190 countries have populations of less than 5.8 million, but 97 percent of the world's people live in countries that are larger than the median country in population. Likewise, most people will live in high-population centuries, and you are likely to be one of them. Indeed, the twentieth and twenty-first centuries have the largest populations of any so far.

Many people think we are special, and lucky, to be born during a remarkable time in which great discoveries such as space travel, atomic energy, and genetic engineering are being made. The Copernican principle says, however, that you are likely to live in a high-population century and, since it is people who make discoveries, you are likely to be born in an interesting century when many discoveries arise. But your chance of being born 200,000 years after the beginning of your intelligent lineage, in the very century when a critical discovery is made that automatically guarantees it, say, a billion-year future, is very small, because a billion years of intelligent observers would be born after such a discovery, and you would be more likely to be one of them.

You are likely to live near a population peak, in an epoch of overpopulation when people have nearly filled their ecological niche. In his book *How Many People Can the Earth Support?* Joel Cohen reports that the median estimate by experts of the maximum carrying capacity of Earth is 12 billion people. Our current population of 6 billion is within a factor of 2 of this number. You will probably live after some event (like the discovery of agriculture) that causes the population to soar, but before some event that causes the population to drop. People who warn of a future population drop should be taken seriously, therefore. Such a population decline might be brought about by an ecological or technological disaster, nuclear or biological war, epidemic, or simply by people choosing to have fewer children. If couples, on average, were to have just 1 child, this could cause the population to drop by a factor of 1,000 in 300 years. Moving from a world with 6 billion people to a world with 6 million people sounds catastrophic, but it might not be any more unpleasant than taking a trip from New Jersey (where the population density is 1,000 people per square mile) to Alaska (where the population density is 1 person per square mile).

Yet such a drop would be dangerous. Species extinction need not be caused by a single event. One event could cause a significant population drop, making a species more vulnerable to the effects of an unrelated event at some later time, which could cause the final extinction. A graph of the population history of our species might therefore show low levels during its initial hunter-gatherer phase, then a brief spike to 12 billion because of civilization, followed by a crash back to hunter-gatherer levels. You expect to live in the spike because most people will. Civilization (with cities and writing) has been around for only 5,500 years, giving 95 percent confidence that its future longevity will be longer than 140 years but less than 214,000 years. Characterized by rapid change, civilization may well be unstable over long periods—slipping out of existence quickly, relative to the species as a whole. Since we have observed only one population spike in human history so far, the Copernican principle tells us there are not likely to be many (meaning more than 39) in the future. We could be living in the one-and-only spike right now.

How many people are likely to be born in the future? The Copernican principle indicates that the chance is 95 percent that we are in the middle 95 percent of the chronological list of human beings now. Since studies of past population indicate that about 70 billion people have been born in the 200,000-year history of our species up to now, we can say with 95 percent confidence that the number born in the future will be at least 1.8 billion but less than 2.7 trillion.

Once, after my *Nature* paper came out, a radio interviewer asked me, "Did you ever ask yourself, 'Aren't I special to have discovered this wonderful thing?'—namely, that the Copernican principle could be used to predict the future?" I had been expecting that question. I told him that I had gone to the Population Research Library at Princeton to look up all the papers I could find on future population estimates. Many papers predicted our population would rise in the next century to about 12 billion before leveling out at about that value and remaining there essentially forever. No one seemed to realize that such scenarios were at variance with the Copernican principle. If there were a brief period of exponential growth followed by a long plateau of high population, then almost everybody would be born in the long, high-population plateau—but you aren't, so that would make you special. I wondered whether anyone else had ever thought of this. The Copernican principle itself nagged at me, reminding me that I should not be so special; others should have thought of it too. But why was I not finding any papers on this? I knew that history is full of famous cases of scientists—including Newton, Darwin, and Copernicus—who discovered important and potentially controversial results but were slow to publish. Besides, there was my own example—I discovered the Copernican argument in 1969, using it to forecast the future of the Berlin Wall. Though I told many friends about it over numerous lunch conversations, I was publishing only after being prodded by the fall of the wall itself. I remember thinking that although I could find no papers on this topic, others had probably thought of it but not published it, at least not prominently enough for me to have found it. That would make me less special.

My reasoning turned out to be correct. When I sent my paper to *Nature*, one of the referees they sent it to was Brandon Carter, the world's foremost expert on the anthropic principle—the idea that intelligent observers must be found in habitable locations in the Universe. Referees are usually anonymous, but they can reveal their identity if they choose, as Brandon Carter did in this case. He gave my paper a rousing endorsement. Carter went on to note that as far as future population was concerned, he had had similar thoughts, namely, that it was unlikely that you should find yourself among the first tiny fraction of all humans to ever live. He had expressed these thoughts at the end of a public lecture on the anthropic principle in 1983, but he had not published it. Later, the noted Canadian philosopher John Leslie heard of Carter's talk, became convinced of the idea and of its importance, and published commentaries on it in the *Bulletin of the Canadian Nuclear Society* in 1989, *The Philosophical Quarterly* in 1990, and *Mind* in 1992. Carter noted that the Danish physicist Holgar Nielsen had also come to similar conclusions about future population in a paper published in *Acta Physica Polonica* in 1989. I was glad to add these references to my paper. I had found some kindred spirits.

Nielsen, without mentioning the Copernican principle explicitly, concluded, like me, that you should expect to be located randomly on the chronological list of human beings. He correctly noted that this meant it was likely for the number of future humans to be of the same order of magnitude as the number of past humans, and it would be unlikely for you to find yourself in the first tiny fraction of all humans to be born. He considered two types of extinction models: (1) sudden extinction, whereby the population rises steadily until it suddenly drops to zero, and (2) a gradual decline, in which the fall-off after the population peak mirrors its rise. In the sudden extinction case, he concluded that the end of the species is close because, with our current, relatively high population, it won't take many centuries for us to accumulate a number of humans in the future of the same order as the number seen in the past. In the gradual decline model, he noted that although we might last as long as we had in the past, the results could still be viewed as terrible: since the population rise in the past has been so rapid, its mirror image in the future would be viewed by many as a catastrophic decline.

I agree that in the sudden extinction model, observers cluster preferentially near the end. In my paper I noted that if population were to rise steadily prior to a sudden end, I would have to revise my 7.8-million-year upper limit downward to only 19,000 years. But I also noted that this is the most pessimistic of all population models. If the population simply drops to a lower level rather than suddenly falling to extinction, the future can be as long as the past. Suppose the fall-off after the population peak mirrors the rise except for being stretched by a scale factor in time. Then the 95 percent confidence level upper limit on the future of the human race would still be 7.8 million years, because if the decline took 39 times longer than the rise, $^{39}/_{40}$ of the people would be born after the peak. Since we have no knowledge of population histories of other intelligent species, it is more conservative to assume that our current population spike simply occurs at some random time in human history—a view that accommodates many possible population scenarios rather than adopting the most pessimistic one (sudden extinction). If you are living in the spike and the spike occurs at some random time in human history, then the 95 percent confidence limits for our future longevity are just as we calculated before: greater than 5,100 years but less than 7.8 million years. (Indeed, the birth of agriculture, initiating the spike, seems to have been facilitated by a random climatological event—the end of an ice age.) In general, use time past to predict time future, and use the number of people in the past to predict the number of people in the future.

Carter and Leslie made their case about future human population from the viewpoint of Bayesian statistics, a rather different statistical treatment, but they also reached similar conclusions. Bayesian statistics, named after the Reverend Thomas Bayes (1702–1761), forms the basis for much of modern probability theory. Bayesian statistics suggests how prior beliefs should be revised upon inspection of new observational data. (Bayes's theorem says that your prior beliefs about the odds in favor of two hypotheses must be revised by multiplying them by the likelihood of your observing what you see, given the two different hypotheses.) This Bayesian shift allowed Carter and Leslie to argue that one is unlikely to find oneself in the first 0.01 percent of all human beings to be born. Precisely how unlikely would depend on one's prior beliefs about future outcomes for the human race. Since we have no prior actuarial data on other intelligent civilizations to aid our calculations, I argue that rather than relying on subjective prior beliefs about the human race, one should instead adopt what is called a *vague Bayesian prior belief*, a prior belief that is properly agnostic about how large the total human population might eventually become—viewing each a priori order-of-magnitude estimate as equally valid and then revising these estimates based on the observational fact that you are approximately the 70 billionth person born. Sir Harold Jeffreys of Cambridge University pioneered this technique in 1939. In a 1994 paper in *Nature*, I was able to show that a Bayesian treatment using Jeffreys's methods gave exactly the same 95 percent confidence level limits as the Copernican results did. It is reasonable that both treatments should agree because they both warn against accepting hypotheses in which what you are observing is unlikely.

THE FUTURE OF THE SPACE PROGRAM

Given these implications of the Copernican principle, let's consider what we could do to improve the survival prospects of our species. Self-sustaining colonies in space would provide us with a life insurance policy against any catastrophes that might occur on Earth, a planet covered with the fossils of extinct species. *The goal of the human spaceflight program should be to increase our survival prospects by colonizing space.*

The Greeks put all their books in the great Alexandrian library. I'm sure they guarded it well, but eventually it burnt down. Fortunately, some copies of Sophocles' plays were stored elsewhere, for these are the only ones of his that survived (7 of 120 plays). Chaos theory tells us that we may be unable to predict today the specific cause of our final demise as a species. By definition, whatever causes us to go extinct will be something the likes of which we have not experienced so far. We simply may not be smart enough to know how best to spend our money on Earth to ensure the greatest chance of survival. We may with good motives spend money saving a certain section of rain forest, only to have that section give rise later to a fatal virus that begins killing us all. But spending money to plant colonies in space gives us more chances—like storing some of Sophocles' plays away from the Alexandrian library.

How long is the human spaceflight program likely to continue? In my May 27, 1993, *Nature* paper I noted that the program was only 32 years old; and I predicted with 95 percent confidence that it would last at least another 10 months but less than another 1,250 years. Since my paper's publication, the human spaceflight program has lasted longer than the 10-month predicted minimum, proving half of my prediction correct already.

Some people figure that even if our interest in space is waning and the space program ends soon, perhaps in the coming century, we will eventually return to space when better technology makes space travel cheap. They liken Neil Armstrong's flight to the Moon to Leif Ericson's trip to North America—a visit several centuries ahead of its time. The Viking effort in America collapsed, but 5 centuries later, Columbus crossed the Atlantic. On this model, we might abandon space travel in the twenty-first century only to resume it again in the twenty-sixth century, with a wave of colonization taking us to Mars and eventually throughout the galaxy over the next billion years.

But the Copernican principle tells us this scenario is not likely. You live in an epoch of space travel now. If two epochs of space travel occur, one short and one long, which one are you likely to find yourself in? The long one, of course! The total number of future years of human spaceflight was likely to be less than 1,250, whether it spanned one continuous period or was divided into several. That's because the year of my paper, 1993, should appear randomly on the chronological list of space-travel years.

Thus, only a relatively brief total epoch of human spaceflight is likely, a brief window of opportunity during which we will have the chance to colonize away from Earth. If we do not succeed in colonizing space during this period, we will be stranded on Earth—subject to all the dangers that routinely cause species to go extinct here.

Since time is short, we should concentrate on establishing the first self-supporting colony in space as soon as possible. That it be self-supporting is important since this would allow it to continue even if funding for future launches from Earth were cut off. Existence of even one self-supporting colony in space might as much as double the long-term survival prospects of our species—by giving us two independent chances instead of one.

We might want to follow the Mars Direct program advocated by American space expert Robert Zubrin. But rather than bring astronauts back from Mars, we might choose to leave them there to multiply, living off indigenous materials. We want them on Mars. That's where they benefit human survivability. Zubrin has shown that a Saturn V–class launch vehicle can deliver a useful payload of 28.6 tons to the Martian surface. According to his calculations, two Saturn V launches could deliver 4 astronauts to the surface of Mars and return them to Earth. By comparison, Gerard O'Neill of Princeton estimated that a self-supporting space colony, with a closed ecology, could be built with a weight of about 50 tons of biosphere per person. Thus, establishing a self-supporting colony of 8 people on Mars might require, at minimum, about 18 Saturn V–class launches—2 for the astronauts, 2 for emergency return craft (providing extra habitat on the surface and with good fortune not used), and 14 to deliver the 400 tons of materials necessary to establish the colony biosphere. This is only slightly more than the 16 Saturn V rockets made in the Apollo program.

Many people might hesitate to sign up for a one-way trip to

Mars. But the beauty is that we only have to find 8 adventurous, willing souls. We just have to find 8 people who would rather spend the rest of their lives exploring Mars and founding a new civilization than returning to a ticker-tape parade in New York City. The colonists' tasks over the next 30 years would be to have 16 children and triple their habitat size using Martian materials. (To ensure genetic diversity, additional frozen egg and sperm cells could always be taken along.) If the colony continued to double in size every 30 years, in 600 years the population could be as large as 8,000,000. In the very long run, as NASA astrogeophysicist Christopher McKay has described, Mars might even be engineered to have a more Earth-like climate and atmosphere, a possibility known as *terraforming*. I'm not saying this would be easy—the Copernican principle suggests it would not—but it's what we should be trying.

Colonies are an incredible bargain. One only has to send a few astronauts. They then multiply at no further cost to us: the colonists do all the work. Colonies can also establish other colonies. After all, the first words spoken on the Moon were in English, not because England sent astronauts to the Moon but because it planted a colony in North America that did. By planting a colony on Mars we may also double our chances of eventually going to the region of Alpha Centauri, because a thousand years from now, who can say whether people from Earth or people from Mars would be more likely to launch the expedition?

Establishing a Mars colony would probably require the human race to spend about as much money in real dollars on human spaceflight in the future as it has in the past, and over a similar time scale—not something unreasonable to ask for. The real space race is whether we will successfully colonize space before the money for space exploration runs out. If we lose that race, we will be stranded on Earth, where we will surely go extinct eventually, probably on a time scale of less than 8 million years.

Massive technological projects often dwindle or die when their underlying causes vanish. In his book *Riddle of the Pyramids*, Kurt Mendelssohn describes the economics of pyramid building and compares it with the space program. The ostensible purpose of the pyramids was to furnish the pharaoh with a tomb. But pyramid building flourished right after the unification of Upper and Lower Egypt into one Egyptian state, when having a big public works project helped bring the country together. In fact, Mendelssohn argues that this was the real reason for their construction. Once the state was well established, this reason vanished. The time from the first pyramid, the 140-foot-tall step pyramid at Saqqara, to the tallest pyramid, the 481-foot-tall pyramid of Cheops, was only about 90 years. After that, smaller and poorer quality pyramids were built until all pyramid construction ceased after about a thousand years. Later pharaohs were simply buried in less expensive valley tombs, like King Tut's.

Although the ostensible purpose for sending men to the Moon was space exploration, its true underlying cause was the Cold War. Space spectaculars, starting with *Sputnik* and Yuri Gagarin's flight, were Khrushchev's way of proving that the U.S.S.R. had the missile technology to deliver nuclear weapons anywhere in the world, without actually using them. Kennedy responded by setting the goal of sending men to the Moon. Since the Cold War has ended, space travel is in danger. On the 25th anniversary of the first lunar landing, during a TV interview on CNBC, I said, "I'm worried that we'll see the day when nobody's left alive who's walked on the Moon." What a sad day that will be—and, to many people, a great surprise. But I suspect it will be greeted with a wistful, nostalgic resignation, rather than a fierce resolve to go back to the Moon and beyond.

People may well say, "How sad, an epoch has passed; what wonderful things we used to do; too bad we can't imagine doing such things today." We would be like latter-day Egyptians looking back in wonder at the ancient pyramids.

In the 1960s the argument was made that traveling to the Moon was too expensive in light of other demands on our resources, such as poverty, Vietnam, civil rights, and other problems, and that we should simply wait until the 1990s, when technology would make it much cheaper. But actually it became much more difficult at century's end to raise money to go to the Moon. Fortunately, we went in the 1960s when we had the chance. If we had waited, we would have missed our chance and would not have visited the Moon yet.

In 1969, Wernher von Braun, chief rocket engineer for the Apollo program, had plans to send humans to Mars by 1982. It didn't happen. Richard Nixon decided not to go to Mars, to end the Apollo program prematurely, and to dismantle the Saturn V assembly line. Confronted with von Braun's plans to go to Mars, he chose to turn away. Three Saturn V rockets that had been built were never launched but were left as museum pieces. The dies for building the Saturn V have been destroyed. This marvelous rocket was allowed to go extinct, with only the smaller shuttle to replace it. In 1989, President Bush promised to send humans to Mars by 2019. Instead of getting closer, Mars is getting farther away. Things do not always become easier to do with the passage of time, and expensive efforts are often abandoned after a while.

In this connection, Timothy Ferris has noted that the fifteenth-century Chinese abruptly abandoned all their naval explorations just after having gone as far as Africa. Or consider another example. In the 1600s Shah Jahan built the Taj Mahal as a tomb for his wife, Mumtaz Mahal, who had died in childbirth. Fashioned of shining white Indian marble, it is, to many who have seen it, including me, the most beautiful building in the world. According to popular legend, Shah Jahan also planned his own tomb: a twin of the glistening Taj but in black marble, facing its sister across the river. Connecting the two would be a dazzling black-and-white inlaid marble bridge. What a sight that would have been! But it was not to be. Shah Jahan's son, Aurangzeb, usurped the throne and put his father under house arrest. A black Taj was never built. The time to build it would have been during Shah Jahan's time—all the artisans were assembled and the expertise and economic means were right there. Of course, people knew the story, and they could have gone back to build it at any later time, but they did not.

If we don't act when we have the chance, that chance may not come again. If we abandon space travel, starting up again may be as difficult as going back and building that black Taj Mahal.

So this is a warning that our species should be getting off Earth and spreading out now, while we have the chance. We observe two frightening facts: our species has not been around very long (only 200,000 years out of a possible 13 billion years), and our species has just a tiny geographical range (one tiny planet out of a vast universe). These are two facts that one might expect to be correlated. Species with a limited geographical range don't last as long as ones with a larger geographical range—simply because the latter are harder to wipe out. Species confined to single islands always face the greatest extinction danger. But in the universe, Earth is just a tiny island. We stay bound to Earth at our peril. And yet this warning is a double warning, for it cautions that the warning itself will likely not be heeded. Why not? Because you were born on Earth. Thus, of all human beings ever to be born—past, present, and future—a significant fraction must be born on Earth, or else you would be special. That means that it is not likely for the human race to heed this warning and escape Earth, moving out in a big way into the vast universe. And this may well be the reason for our likely early demise as a species. Abandoning the human space-flight program would be a tragic mistake, yet it is a mistake we are likely to make.

THE TIME TRAVELER'S LESSON

Time travel is a project for supercivilizations. Time travel to the future requires a civilization already accustomed to interstellar travel. Time travel to the past could be attempted by supercivilizations commanding the energy resources of an entire galaxy. Perhaps a billion habitable planets reside in our galaxy. A supercivilization that had colonized its entire galaxy could have a population a billion times larger than we have on Earth today. Such supercivilizations must be a billion times rarer than civilizations confined to their home planet, or else they would dominate the number of intelligent observers in the universe, and you would likely find yourself living in such a supercivilization.

You are an intelligent observer—someone who is conscious and able to reason abstractly. As far as we know, our species is the first species on Earth to qualify as intelligent observers. Chimpanzees and porpoises, cockroaches and bacteria don't ask questions like "How long will my species last?"

As an intelligent observer, your location in our universe must be special to the extent that it must be among the subset of habitable locations. This is the key insight of the weak anthropic principle, as formulated by Brandon Carter in 1974, a line of reasoning first applied by Princeton professor Robert Dicke in 1961. Dicke reasoned that as an intelligent observer, you were likely to find yourself about 10 billion years—one stellar lifetime—after the big bang. Much before that, the stars wouldn't have had enough time to produce the carbon necessary to make everyone from cooks to physicists. Much after that, stars would have burnt out, and the universe would be much less habitable. My application of the Copernican principle recognizes that you may be at a special epoch in our universe precisely because you are an intelligent observer, but *among* those intelligent observers, you should not be special. You should expect to be randomly located on the chronological list of intelligent observers in our universe. Furthermore, you should expect to live in an epoch of the universe in which the population of intelligent observers is high because most intelligent observers would live in such an epoch. If intelligent civilizations typically lasted forever, then almost all intelligent observers would live in the far future, long after the stars had burnt out. This does not imply that there must be no intelligent life in the far future, just that a significant fraction of all intelligent life occurs at the present star-burning epoch, when our universe is most habitable. There could be some intelligent life forms in the far future; they just must be rare. Otherwise, you would likely be one of them. Likewise for supercivilizations, the Copernican argument does not imply there are none—they simply must be rare.

At lunch one day in Los Alamos in 1950, the noted physicist Enrico Fermi asked a famous question about extraterrestrials: "Where are they?" The answer to Fermi's question, provided by the Copernican principle, is that a significant fraction of all intelligent observers must still be sitting on their home planet, just like you; otherwise, you would be special. Simple.

If you think space colonization occurs frequently, you should ask yourself, "Why am I not a space colonist?" If you think most intelligent observers in the universe are intelligent, self-conscious computers, or genetically engineered beings, you should ask yourself, "Why am I not an intelligent computer? Why am I not genetically engineered?"

The universe is a big, perhaps infinite, place, and the occasional intelligent species could be much more successful than we are now, but most are probably not. The Copernican principle says you are likely to come from an intelligent species having a population now larger than the median. This is true for the same reason that you are likely to come from a country having a population larger than the median: simply because most intelligent observers will. Thus, in terms of population, we are right now likely to be one of the larger, more successful intelligent species.

The fraction of civilizations like ours that eventually turn into supercivilizations with their enormous populations must be extremely low; otherwise, you would likely be sitting in one right now. Supercivilizations might indeed do amazing things, but we are not likely to become one of them. Some intelligent species might develop time travel to visit the far future or even the past, but probably most do not. Time travel is difficult. If almost all intelligent observers in the universe did it, since you don't, that would make you special. That doesn't mean that time travel is impossible, just that it must be rare at best. As Darwin pointed out, most species do not achieve their maximum potential. Some fish lay a million eggs—but most of those eggs don't develop into adults. Similarly, most species leave no descendant species. Things don't usually work out as well as they conceivably could. This is precisely the reason so many people wish they could visit the past, to change things that have gone wrong—from saving a lost loved one to stopping Hitler before he came to power. Life is often tragic.

Intelligence offers the possibility of immense power and longevity, but this potential must be fully realized only occasionally—otherwise our situation would be very atypical. The moral is both exhilarating and distressing.

From this perspective, intelligent life is potent in principle but, being complex, usually fragile in practice. We've amassed a track record of only 200,000 years on one tiny speck in this vast universe that's already 13 billion years old. We are not very powerful—we control sources of energy that are tiny even compared to the Sun. And we do not enjoy a long past longevity.

Humbling as these facts may be, in this short period we have also done something remarkable. We have figured out a great deal about the laws of physics and the universe. We know the universe was much smaller in the past than it is now, we have some idea about how galaxies formed and how Earth got here, and we're smart enough to have discovered where we are in the universe. This level of understanding is remarkable. And if we understand these things, then a reasonable fraction of all intelligent observers must understand them also. But it's precisely in feats of understanding, rather than in longevity or power, that one might expect intelligent observers to excel. The ability to ask questions seems to give some ability to answer them, but it doesn't give us a lot of time. That is the essence of the report from the future. One of the things we should understand about time is that we have just a little.

Don't waste your time, humanity. You have just a little. It is the time traveler's secret.

NOTES

ANNOTATED REFERENCES

INDEX

NOTES

1. Dreaming of Time Travel

21 *Somewhere in Time*: This is based on Richard Matheson's book *Bid Time Return* (New York: Viking Press, 1975).

His world line is indeed complex: Michio Kaku includes a spacetime drawing of "Jane's" world line in his 1994 book, *Hyperspace* (New York: Doubleday), p. 241.

30 *Quantum mechanics:* Black hole physics also had its share of paradoxes. Jacob Bekenstein proved that there was a finite amount of disorder (called entropy) associated with a black hole. Stephen Hawking and others showed that for this to be consistent with the laws of thermodynamics, the black hole would have to exist at a finite temperature. But that made no sense—all objects at finite temperature give off thermal radiation, while black holes could not give off any radiation at all; no radiation can escape from them. It was a true paradox. Then Hawking came up with a quantum effect that would cause a black hole to emit radiation. This has been named Hawking radiation and is Hawking's greatest discovery. So wherever paradoxes lurk, there is a chance for some great physics to emerge.

Stephen Hawking thought of a different way: Along with Simon and his colleagues and Stephen Hawking, Seth Rosenberg at the University of California at Santa Barbara and Arley Anderson at Imperial College in London have tackled this general problem of calculating quantum probabilities in the presence of time machines. All use different methods and arrive at different answers, so work on this problem continues.

2. Time Travel to the Future

39 *Maxwell knew the velocity of light:* Since I will be referring to

the speed of light through empty space often, I have rounded it to an even 300,000 kilometers per second. Its actual speed through empty space is, more precisely, 299,792.458 kilometers per second.

42 *The claim that they were at rest:* Here's an example of observers moving at a uniform velocity and their experience of being "at rest." On a plane trip, have you ever noticed that once the craft has achieved its cruising altitude and is traveling smoothly at a constant speed (without turning), it feels just as if you are on the ground? You can balance a coin on your tray table or walk up and down the aisle just as you would if the plane were sitting on the runway. In fact, if all the window shades in the plane were pulled down so you couldn't look out, you would have a difficult time telling whether you were on the ground or barreling along in the air at 500 miles per hour. The only clues might be auditory (engine sounds, whistling wind), but you would not feel any difference between sitting on the runway and flying along.

44 *On board is an astronaut:* The astronaut and I can check that our two light clocks have the same distance between their mirrors by doing a clever test, presented in a slightly different form by E. F. Taylor and John A. Wheeler in *Spacetime Physics* (San Francisco: W. H. Freeman, 1992). We line up our clocks perpendicular to the direction in which the astronaut is flying past. For example, if the astronaut is passing me from left to right, we can position our clocks vertically, so the light beams go up and down. Let the astronaut mount his clock on the outside of his rocket and fly close enough for the two mirrors of his clock to make scratches on the wall of my laboratory as they pass by. Likewise, let me put the two mirrors of my clock just outside my lab so they can make scratches in the side of the astronaut's rocket as it goes by.

Suppose I were to observe that the scratches on my laboratory wall were separated by less than 3 feet so they both lay between the two mirrors of my vertical clock. Then the astronaut would have to observe that my mirrors passed outside his to make scratches on the side of his rocket that were wider than the separation of his mirrors. We would both agree that my clock was taller than his. I, believing myself at rest, would think that measuring sticks carried by a rapidly moving observer were somehow always shortened in the direction perpendicular to the line of motion. He, by contrast, would think he was at rest and would conclude that measuring sticks carried by a rapidly moving observer (me) were always lengthened in the direction perpendicular to the line of motion. But that would violate the first postulate because it would mean that the laws of physics looked different to me and the astronaut. That's not allowed.

A similar problem would arise if our roles were reversed and the scratches the astronaut's light clock mirrors made on my lab wall were wider than the distance between my mirrors. The only way we would both see the same physical effects would be if my clock's mirrors made scratches on the side of his rocket that were 3 feet wide as measured by him and his clock's mirrors made scratches on the side of my laboratory wall that were 3 feet wide as measured by me. That is, the two sets of mirrors could scratch each other as they passed. Then both our observations would be the same, as demanded by the first postulate. This ensures that his measuring rods and mine are measuring the same thing. Einstein took nothing for granted.

55 *All observers agree on the quantity:* The quantity different observers can agree on is called ds^2. We write $ds^2 = -dt^2 + dx^2 + dy^2 + dz^2$, where dt means the difference in time between two nearby events, dx is the difference in the left-right direction, dy is the difference in the front-back direction, and dz is the difference in the up-down direction. Note the minus sign on the term for the time dimension, which distinguishes it from the three spatial terms.

Alpha Centauri: Since I refer to it often, I have conveniently rounded the distance to Alpha Centauri to the nearest light-year (4 light-years). What we call Alpha Centauri, the star system nearest to the Sun, is actually a triple star system: Alpha Centauri A, a solar-type star; Alpha Centauri B, a lower-luminosity, orange-colored star; and Alpha Centauri C, a very faint red-dwarf star. A and B form a binary system about 4.35 light-years from Earth. Alpha Centauri C (sometimes called Proxima Centauri) is currently a bit closer—4.22 light-years from Earth. When people speak of Alpha Centauri, they are usually referring to A, the solar-type star.

As an amateur astronomer in high school, I always longed to observe Alpha Centauri, but because it is a south circumpolar star, from my home in Kentucky it was always below the horizon. I saw it for the first time years later from Tahiti while on a trip around the world. When I got to Tanzania, I was able to observe both Alpha Centauri A and B through a small telescope—a true thrill for me.

60 *Flatland:* In Flatland, we would write $ds^2 = -dt^2 + dx^2 + dy^2$ because there would be only two dimensions of space.

61 *Lineland:* In Lineland, we would write $ds^2 = -dt^2 + dx^2$.

62 *dreamtime:* If we had two dimensions of time (time and dreamtime) in addition to the three spatial dimensions, we would write $ds^2 = -dt^2 - dd^2 + dx^2 + dy^2 + dz^2$, where dd means difference in dreamtime between two events. Note the minus sign associated with both the time and dreamtime dimensions.

3. TIME TRAVEL TO THE PAST

83 *making a cylinder:* I've used this method of illustration on TV, and Igor Novikov also depicted it in his book *The River of Time* (Cambridge, England: Cambridge University Press, 1998).

87 *how Einstein's equations look:* The terms in Einstein's equation have two indices, indicated by the subscripts, which, in a four-dimensional spacetime, can each take on one of four values, for the time dimension and the three spatial dimensions; hence, this one equation really stands for $4 \times 4 = 16$ equations. That's why I can write this one equation and still properly refer to Einstein's *equations*. Fortunately, some of these equations are automatically equivalent to each other, so we are left with 10 independent equations we must solve.

88 *Einstein said of his travails:* This quotation is from C. W. Misner, K. S. Thorne, and J. A. Wheeler, *Gravitation* (San Francisco: Freeman, 1973), p. 43, citing M. Klein as the source.

90 *Kurt Gödel:* Gödel was already famous for his incompleteness theorem, published in 1931. Prior to Gödel, mathematicians had hoped to find a finite system of axioms that would pave the way for proofs of all true theorems in the field. Working in such a system, Gödel proposed a self-referencing theorem, which in lay terms can be stated as: THIS THEOREM IS UNPROVABLE. Suppose you can prove this theorem; then the theorem is false, and that's a problem, because no good set of axioms should allow you to prove a theorem that is false. But suppose you can't prove the theorem; then the theorem is true, but you can't prove it using your axioms. Either way, the axioms fail

to achieve their goal. Mathematics is incomplete. Gödel's theorem is perhaps the single most important development in mathematics in the twentieth century.

94 *William Hiscock:* Mathematically, Vilenkin's approximate solution for a cosmic string looks like this: $ds^2 = -dt^2 + dr^2 + (1 - 8\mu)r^2 d\phi^2 + dz^2$. Now compare the exact solution found by me and by Hiscock: $ds^2 = -dt^2 + dr^2 + (1 - 4\mu)^2 r^2 d\phi^2 + dz^2$. Only a small difference! ds^2 is a quantity different observers can agree on, dt is the difference in time between two nearby events, dr is their difference in radial distance r from the string, $d\phi$ is their difference in angle around the string, and dz is their difference in vertical distance up and down the string. μ is the mass per unit length in the string in units of Planck masses (2 × 10^{-5} grams) per Planck length (1.6×10^{-33} cm) = 1.25×10^{28} grams/cm. Since we might expect a value of $\mu \sim 10^{-6}$, the approximate solution is quite close to the exact solution.

104 *To allow time travel to the past:* The speed required to produce time travel depends on the mass per unit length in the strings. The less massive the strings are, the smaller the missing wedges they create, and the smaller the shortcuts, so the faster the strings are required to move to effect time travel. But for any particular mass per unit length, we can always find the appropriate speed (slower than the speed of light) for the strings to pass each other to enable time travel to the past.

106 *Guth and his two MIT colleagues:* String solutions, as I have mentioned, are related to solutions involving masses in Flatland— just eliminate one spatial dimension. Carroll, Farhi, and Guth's papers, together with a result by Gerard 't Hooft, *Classical and Quantum Gravity* 9 (1992): 1335, showed that in Flatland, with static initial conditions or ones with slowly moving masses, one could not construct a time machine (assuming only positive masses were allowed). Of course, my time-travel solution, with two masses moving at nearly the speed of light, did not have such initial conditions; so their papers posited a rather restricted set of initial conditions. Our own universe started off with a rapid expansion—the big bang—so Matthew Headrick and I argued in a 1994 paper that one does not want to adopt undue restrictions on initial conditions. Part of our argument was that in Flatland one cannot associate rotations (and kicks in velocity),

incurred after circling masses, with momentum, both because mathematically such rotations do not add as momentum is supposed to do and because these spacetimes at large distances are not approximately flat—so defining a momentum is not possible for them in any case. Furthermore, that momentum analogy wouldn't translate to four-dimensional spacetimes.

116 *tidal forces might not tear you apart:* Hawking radiation, a quantum process causing the black hole to eventually evaporate (in 4×10^{94} years for a 3-billion-solar-mass black hole), adds complications, altering the geometry and limiting the entrance of very late photons.

117 *As Kip Thorne says:* K. S. Thorne, *Black Holes and Time Warps* (New York: Norton, 1994), p. 479.

the argument has a loophole: My cosmic string solution, with two infinite cosmic strings, avoids Tipler's theorem because its Cauchy horizon extends to infinity, and therefore it has no singularities on it. Curiously, the unperturbed rotating black hole solution avoids Tipler's theorem as well. Since this black hole solution lasts forever, the Cauchy horizon goes on forever even though it is curled up in a finite region. It has no singularities. There is a ring singularity, but it occurs later. The time traveler sees it only once she has crossed the Cauchy horizon. Again, she might be killed by particles emitted unpredictably by the ring singularity, but then again—remembering the big bang—she might not. If the black hole evaporates through Hawking radiation, as we expect, the black hole will not last forever, and a singularity may indeed intrude on the Cauchy horizon.

124 *travel among the stars:* The *Enterprise*'s 5-year mission was to explore a new star system each week and report back to Star Fleet Headquarters. The *Enterprise* could have visited a new star system each week (as measured by clocks aboard the ship) simply by traveling at 99.999 percent of the speed of light, assuming the stars are 4 lightyears apart. The crew would age slowly because the ship was moving at nearly the speed of light. But when 5 years had elapsed according to the crew, they would find on their return to Star Fleet Headquarters that it was over 1,000 years later according to people there. To report back to Star Fleet Headquarters within 5 years of Star Fleet time, after visiting many star systems, would require faster-than-light travel.

125 *Imagine yourself as an ant:* Kip Thorne has used the meta-

phor of ants crawling on a rubber sheet to explain black holes in *Scientific American* 217, no. 5 (1967): 96 and in *Black Holes and Time Warps*, p. 247.

126 *two moving warpdrive shortcuts:* If you could create a warpdrive path that would allow a starship to get to Alpha Centauri in a few minutes, then your trip would connect two events separated by a greater distance in space than in time. Thus, as in the cosmic string case, an observer in a rocket ship traveling at a certain speed would view your departure from Earth and your arrival at Alpha Centauri as simultaneous events. The spacetime outside the narrow path taken by the starship (the slit) would be unperturbed. If that rocket observer saw you leaving Earth at noon and arriving at noon on Alpha Centauri the same day, then you could make a second warpdrive—a second slit—going from Alpha Centauri to Earth that would allow you to depart Alpha Centauri at noon and return to Earth at noon, according to him. You could then return to shake hands with yourself as you departed.

127 *nothing can travel faster than light:* You may have learned of a story picked up by the media that someone had outrun a light beam in the lab. Such attempts usually involve quantum tunneling. If you sit on one side of a wall, a small probability exists that you will tunnel through and suddenly find yourself on the other side. Since—if you do tunnel—you effectively go from one side of the wall to the other faster than light, you can beat a light beam on a parallel path. For example, in his lab, Raymond Chiao of the University of California at Berkeley set up a race between photons that traveled on a path straight to a detector and others that tunneled through an opaque sheet of glass a few hundred thousandths of an inch thick. Photons that tunneled beat those that didn't by an average of 1.5 quadrillionths (1.5×10^{-15}) of a second.

But a complication arises. When one sends photons on a race in the lab, they are contained in a wave packet of finite length. Imagine two packs of runners sprinting at the speed of light; in each pack, some are slightly ahead and others trail slightly behind. One pack goes straight, while the other pack runs into a wall. A tiny fraction of the runners hitting the wall tunnel through it. The runners who do so form a tight group arriving ahead, on average, of the runners going straight. But the leading edge of these two groups of runners is the same, so, it could be argued, you have not really beaten light after all. Since curved spacetime allows one to beat a light beam easily by appreciable amounts (in the case of the gravitationally lensed quasar 0957, one light beam beat another by 417 days), using curved spacetime for traveling back in time seems more promising than attempting quantum tunneling—to say nothing of the problem that, most of the time, one would fail to tunnel.

127 *A tachyon would have to be accompanied by gravitational waves:* These waves accompanying a tachyon are called *gravitational Cherenkov radiation.*

128 *Therefore, tachyons could not be used:* Once in 1973, during a discussion I had about tachyons with Richard Feynman at Caltech, he said that he doubted they would ever be discovered.

4. TIME TRAVEL AND THE BEGINNING OF THE UNIVERSE

132 *the fellows retire upstairs to drink port:* During one such discussion involving Halley's comet and how it returned every three quarters of a century, Mr. Nicholas, a senior fellow then 87 years old, remarked that not only had he seen the comet on its previous appearance in 1910 but that as a young fellow himself he had talked to a then senior Trinity fellow who had seen the comet the time before that—in 1835. Nicholas lived to regale the fellows with stories on his hundredth birthday and walked upstairs to sample the port on that day as well.

136 *the Casimir vacuum at least creates the possibility:* In this type of wormhole (with electrically charged Casimir plates), an astronaut in the middle of the tunnel would age less than either wormhole mouth because, like the stay-at-home time traveler, she would be at the bottom of a deep gravitational well. The two mouths stand still with respect to each other, as seen by her, and the gravitational well is equally deep on both sides, so she sees clocks in both mouths ticking at the same rate—and synchronized. Move one mouth on a round trip near Earth so its clock ticks slowly as seen by earthlings, while leaving the other fixed at Alpha Centauri. The two mouth clocks, which are synchronized as seen through the tunnel, will then connect different times on Earth and Alpha Centauri in the external space-

time, just as we discussed in Chapter 3 except that here the astronaut one finds at the tunnel's center is younger than expected because of the gravitational wells.

That is what's inside a cosmic string: Outside the cosmic string is the normal vacuum, but trapped inside the string is a high-energy vacuum state that could arise as a decay product of an inflationary vacuum state that originally permeated all of space. The strings would then be left fossilized remnants, much like isolated snowmen are left standing long after the snow on lawns has melted.

140 *That football game would always have the same outcome:* In his imaginative tour de force *Einstein's Dreams*, Alan Lightman considers a Groundhog Day–type spacetime in which all the inhabitants are jinn whose world lines circle the spacetime once. These jinn experience the same events repeatedly, giving some an incredible sense of déjà vu. Strictly speaking, the movie *Groundhog Day* visits a Groundhog Day spacetime in the many-worlds picture of quantum mechanics because each time Bill Murray's character returns to the past, he can make different decisions about how he will spend Groundhog Day.

the normal wrapped vacuum in Misner space: The Hiscock and Konkowski calculation of the normal vacuum in Misner space reminded me of the black hole case in which the Boulware vacuum (see next note for definition) blew up as the event horizon was approached. This problem was cured by the introduction of Hawking radiation. I thought a similar Hawking radiation remedy might exist for the blowup as the Cauchy horizon in a time machine was approached.

Since Li-Xin Li found, in agreement with Hiscock and Konkowski, that the vacuum energy density in the time-travel region of Misner space was already positive, it looked as if one could not forestall its blowup as it approached the Cauchy horizon by adding radiation to the solution, as was done with Hawking radiation in the black hole case.

the vacuum state measured by accelerated observers: An astronaut in a rocket in interstellar space accelerating with 1g acceleration, like that experienced on Earth, will see Unruh radiation (photons) with a wavelength of about 1 light-year. He would see the Rindler vacuum becoming increasingly negative farther and farther in his wake, finally blowing up to an infinitely negative state about 1 light-year behind his ship. This is okay, because at that same spot he would deduce an infinite amount of Unruh radiation of infinite positive energy density. The two infinities would cancel each other out to give a total energy density of zero—that of the normal vacuum. This would be akin to having an infinite bank account and an infinite debt—one would still be broke. If the accelerating observer did not detect any radiation, he would be living in a world with a pure Rindler vacuum and no thermal radiation. That world would have truly a negative total energy density, which would blow up (become infinitely negative) at about 1 light-year behind his rocket. Negative energy density causes spacetime to curve, according to general relativity, and an infinite negative energy density would cause a singularity in the curvature of spacetime. The Rindler vacuum is calculated assuming a flat spacetime geometry; if it blows up and changes that geometry, the calculation is no longer self-consistent. Thus, a pure Rindler vacuum in flat spacetime, unaccompanied by any radiation, is not a self-consistent vacuum state. The normal vacuum state is also calculated assuming a flat spacetime geometry, but it has zero total energy density and zero total pressure and so, according to Einstein's equations of general relativity, it produces a flat spacetime geometry. Thus, the normal vacuum in flat spacetime is self-consistent. We are always looking for such solutions. Given a background geometry, if we have a choice of quantum vacuum states, we should pick the self-consistent one, which produces the geometry in which it lives.

For the curved spacetime outside a cold neutron star, physicist David Boulware found a vacuum state now called, not surprisingly, the *Boulware vacuum*, in which external observers would see no radiation. It has a small negative energy density with a finite value even at the neutron star surface—not large enough to significantly perturb the geometry—so it works as a solution. But is the Boulware vacuum right for a black hole? If the Boulware vacuum existed around a black hole, external observers would observe no radiation. Unfortunately, its energy density becomes ever more negative as the event horizon is approached, blowing up—becoming infinitely negative—at the horizon itself. Since that would cause the background geometry to change significantly, this is not a self-consistent solution for the black hole.

Stephen Hawking and his colleague James Hartle discovered, however, that there was another vacuum state for the geometry around a black hole—called the *Hartle-Hawking vacuum*. This vacuum state has only a tiny, finite energy density at the event horizon of the black hole—it does not blow up there. Since it does not significantly perturb the geometry, it can be considered a self-consistent solution. Just as the normal vacuum looks to accelerated observers like a Rindler vacuum plus thermal radiation, the Hartle-Hawking vacuum looks to accelerated observers like a Boulware vacuum plus thermal radiation. Therefore, accelerated observers, firing their rockets to hover above the hole outside the event horizon at a constant distance, will observe thermal radiation—*Hawking radiation*. The part of the Hawking radiation that happens to be directed radially outward becomes greatly redshifted as it climbs out of the black hole's gravitational well, and external observers can see it at great distances. The tiny negative energy density of the Hartle-Hawking vacuum state causes a slow "back reaction," which leads the black hole to slowly lose mass until it eventually evaporates to nothing. The Hawking radiation energy seen by the external observers ultimately comes from the energy the hole loses as its mass decreases. Energy is conserved. Hawking radiation emitted by black holes of seven solar masses and greater is feeble—far beyond our current ability to detect it—but physicists have little doubt that it is emitted.

142 *This wrapped Rindler vacuum:* The Hiscock and Konkowski vacuum—a wrapped normal vacuum—was inconsistent; it did not produce the geometry it started with. Li-Xin Li's wrapped Rindler vacuum was the correct, self-consistent vacuum for Misner space, when the walls approach at 99.9993 percent of the speed of light— just as the Hartle-Hawking vacuum was correct for the black hole.

If one had been as skeptical about the existence of the insides of black holes as Hawking was about the existence of time machines, the Boulware vacuum might have been perceived as the correct vacuum for black holes. After all, that vacuum state produced no radiation, and originally everyone thought that black holes emitted no radiation. (That's why they were called black.) But the Boulware vacuum blew up—developed a negative infinity—as one approached the event horizon of the black hole, changing the geometry and causing the solution to break down before the black hole could be entered. This might have been interpreted as evidence for a "black hole protection conjecture": that quantum vacuum effects always conspired to prevent you from entering the event horizon of a black hole. Hawking, of course, believed in black holes and that you could get inside one, and the correct solution—the Hartle-Hawking vacuum—was found.

142 *Li-Xin Li would be first author:* This paper, "Self-Consistent Vacuum for Misner Space, and the Chronology Protection Conjecture," also discusses difficulties the time traveler would face: avoiding hitting himself and the perturbations produced by the fact that his gravitational field would wrap around the spacetime as well. These problems could be overcome, however, if the time traveler navigated appropriately and took along some of that marvelous negative-energy-density stuff so that the spaceship's total mass would be zero. Then he would not perturb the solution. (Similarly, for a time traveler circling my infinite cosmic strings, we were able to show, using some mathematical results by J. D. E. Grant, that a positive-mass spaceship's gravitational field circling the strings could eventually lead to formation of a black hole—something that already occurs in the finite loop case.)

143 *Our Misner space paper:* Scientific investigations continue. After our papers on Misner space and on time travel in the early universe appeared, Li-Xin Li discovered an improved renormalization procedure for Misner space. Whenever such quantum calculations are done, one obtains infinite answers that must be "renormalized" to give the answers actually observed—by subtracting off the infinite answer from a vacuum state known to have a zero energy density and pressure. For time-travel solutions two techniques are used, the *Euclidean section* method invented by Hawking, and the *covering space* method.

In the first method one solves the problem by treating all the dimensions as spacelike, and in the other, one effectively treats time travelers returning to the past as clones of the original person. These two techniques should give the same results and get around any questions of how to do quantum mechanics in the presence of closed timelike curves. In our Misner space paper we worked the problem using the covering space method, and checked with the Euclidean section method to see if the results were correct—they were. In a later paper in *Physical Review D* 59 (1999): 084016, Li-Xin Li showed that an improved renormalization procedure was needed for the covering space method. With this new procedure, the Euclidean section method and the covering space method yield the same results in the general case. This allowed Li-Xin Li to show that our original results —the finding of a self-consistent wrapped Rindler vacuum of zero energy density and zero pressure for Misner space in which the walls approached at 99.9993 percent of the speed of light—could be extended from the conformally invariant scalar field we had already analyzed to include other fields such as the electromagnetic field.

Hiscock later calculated the renormalized energy density for this case using the *old* renormalization procedure with self-interacting scalar fields, such as those encountered in inflation, and found a blowup. Li-Xin Li and I repeated these calculations using Li-Xin Li's new renormalization procedure and found that self-interacting scalar fields gave zero energy density and pressure as expected. Thus, our results could be extended to these fields as well. Our results for the early universe—including our temperature and entropy calculations—are unaffected by the new renormalization procedure. In superstring theory, renormalization occurs differently, but for each particle there is a supersymmetric partner; for high-energy conditions like those prevailing in the early universe, the quantum effects of the particles and their supersymmetric partners cancel each other out in pairs to prevent quantum blowups in any case. More studies along these lines will surely follow.

168 *no paradox:* Light moves around de Sitter space at the speed of light, of course, but as the space itself begins to expand at nearly that same speed, the light begins to make less and less angular progress around the circumference.

Light signals in the de Sitter spacetime diagram (Figure 21) have world lines that are straight lines tilted at 45 degrees to the vertical (moving horizontally 1 light-year in space for every year upward in time). These lines lie in the curved, hourglass-shaped surface. In fact, you can make such a surface by taking two hula hoops, joining them with a dozen longish pieces of string connected at corresponding points to make a drum shape, and then rotating the top hoop until the strings are all tilted to 45 degrees. You will see an hourglass shape with a narrow waist made of string.

A light beam emitted at the waist will progress by only 90 degrees around the circumference in an infinite amount of time. The circle representing the circumference eventually grows to infinite size, so even with an infinite amount of time, the light beam cannot get all the way around it. If two people live at opposite poles of de Sitter spacetime, they will not be able to exchange radio messages. In fact, any two observers, traveling on geodesics, will eventually find themselves drifting apart ever faster as the space between them expands. Eventually, when by their own clocks they are moving apart at more than the speed of light, they will lose contact with each other. The light signal sent by the first never catches up with the friend. The friend is going slower than the speed of light but has too much of a head start. Although the light beam is going faster, the friend continues to move closer and closer to the speed of light and so always keeps a lead and is never caught. So if you live in de Sitter spacetime, you are always losing contact with your friends. If your friend Fred lives some distance from you, you will see a red shift in the radiation you receive from him as he moves away. Eventually, as the distance between you begins to open up as fast as the speed of light, as measured by your clocks, that red shift will become infinite. His radio message will c-o-m-e e-v-e-r m-o-r-e s—l—o—w——l———y. The last sentence you receive will take an infinite amount of time to arrive. And the sentence he sends after that will never arrive. That signal will be forever passing between him and you but never arriving. It would look to you as if Fred had fallen into a black hole.

171 *An inflationary epoch could provide the bounce:* Physicist Lee Smolin of Penn State and Russian physicist Valeri Frolov and his colleagues M. A. Markov and Viatcheslav Mukhanov have proposed that any time a gravitational collapse proceeds toward the formation of a singularity, such as in a black hole, at the last moment, as the temperature rises, one enters an inflationary vacuum state, which undergoes a de Sitter bounce to form a new inflationary universe. Thus, new inflating universes could bud from our own universe like branches off a tree.

You can't start with nothing: If you put a bit of inflationary vacuum in a box and expanded that box to a larger size, you would have to expend energy to move the walls outward because the inflationary vacuum's negative pressure, or suction, would pull the walls inward, and you would need to overcome that. When finished, you would have a larger box filled with inflationary vacuum. This would have the same energy density as before, but with a larger volume the box would have a greater total energy. The added energy should equal the energy you expended in moving the walls of the box outward. In general relativity local energy is conserved in little regions as expected. But in the entire solution, because space and time curve, the total energy content within the universe is not conserved—there is no flat place on which to stand to set an energy standard. This is a peculiar and important property of general relativity. Imagine the inflating universe divided into many small boxes. Inside each box, an observer would observe that the total energy in her box goes up as her box expands. She would attribute this to someone pulling outward on the sides of her box. But actually what's pulling on the sides of her box are just the adjacent boxes, which are also growing. So, in this case, the total energy content within the entire universe goes up with time as the volume of the universe increases.

173 *The hyperbola shows the surface in spacetime:* Why does the hyperbolic surface have a negative curvature? The brilliant German mathematician Johann Karl Friedrich Gauss showed that a sphere—the set of points equidistant from a central point in space—has a positive curvature. The degree of curvature depends on the sphere's size. A small sphere such as a mustard seed is sharply curved, a bigger sphere, such as a beachball, curves more gently, and a huge sphere like Earth has a curvature so slight that to us it seems almost flat. Gauss found that the amount of curvature is inversely proportional to the square of the sphere's radius. This radius is the distance in *space* between any point on the sphere's surface and its center. By contrast, the hyperbolic surface in our example has a negative curvature because it represents the set of events equidistant in *time*, measured by rockets, from a single event. As we discussed earlier, in special relativity, observers agree on the square of the distance in space minus the square of the distance in time. That minus sign associated with the square of the distance in time gives the hyperbolic surface its negative curvature.

176 *bubbles . . . were the answer:* For my model to work, the high-

density vacuum state needed to remain inside the bubble for a little while after the bubble formed, before decaying into thermal radiation. This would allow the universe to inflate to a size large enough to agree with observations. I needed inflation to continue within the bubble (until those alarm clocks went off at one o'clock in Figure 22) for a period at least 100 times as long as the time it took the outside de Sitter space to double in size according to observers there. This would create a noticeably negatively curved universe. If one let inflation within the bubble go on longer, say by 10 times, then the universe would remain negatively curved but inflate to such large size that it would be indistinguishable from flat today.

182 *The squares of distances:* In the black tunneling region, one is "in" the tunnel and therefore "below ground" in our landscape analogy. Being "below ground" flips the negative sign connected with the time dimension into a positive sign. (We would write $ds^2 = +dt^2 + dx^2 + dy^2 + dz^2$ in each small neighborhood.) Thus, the dimension of time *becomes* a dimension of space just like the other three, giving us four dimensions of space in the black region.

186 *chaotic inflation:* To understand the mechanism of Linde's chaotic inflation, recall our analogy of the bowling ball rolling on a varied landscape. Higher altitude corresponds to higher vacuum energy density and more rapid inflation. Start the bowling ball in the coastal plain. A small chance exists that it will quantum-jump up into the mountains. Once there, rapid inflation will occur, expanding the region to enormous size and creating a baby inflationary universe. Little pieces of this region will lose contact with one another as the baby universe inflates, and these will start behaving independently, like many bowling balls. Most will roll downhill, but occasionally one of these pieces (bowling balls) will quantum-jump even higher in the mountains where it will expand even faster than all the others. That makes a second-generation baby universe whose volume soon exceeds that of all the others because of its more rapid inflation. This process repeats itself. Soon, most of the volume of the Universe exists in pieces that are higher and higher in the mountains, with the inflation largely taking place at the Planck density (5×10^{93} grams/cm^3). These pieces are continually spawning baby universes, populating the mountains with more and more bowling balls, which are continually

rolling down (and occasionally quantum-jumping back up—spawning even more bowling balls). As individual bowling balls roll down, they may roll into mountain valleys and become trapped there. They can then tunnel out of the valleys and roll down, spawning open bubble universes as in Figure 22. Alternately, they may not encounter any valleys and simply roll slowly down the mountain, allowing the whole irregularly shaped region to turn into an enormous (and therefore approximately flat-looking) Friedmann universe.

Linde's chaotic inflation model explains how inflation may occur under very general circumstances. Until we have a theory-of-everything, we don't know what the landscape governing all of this would look like, but Linde's work indicates that quantum fluctuations or quantum jumps should soon lead to baby universes spawning other baby universes in the "mountains," and these would eventually tunnel or roll down to make universes like ours. This process yields a fractal tree of baby universes branching off other baby universes.

190 *if the time loop is short:* If the normal vacuum is stable against the spontaneous formation of cosmic strings (which it is—otherwise cosmic strings would be forming in your bedroom), then our self-consistent quantum vacuum state solution should be stable as well. This follows from adapting an argument by Michael J. Cassidy, one of Hawking's students.

Hawking had noted a problem with time machines not relying on negative-energy-density stuff. An instability exists as one *enters* the time machine at its Cauchy horizon. If one adds a wave of small amplitude to the solution, the wave can travel back in time to its starting point, arriving with more energy than it had originally. It keeps coming around, again and again, growing in energy each time, causing the solution to be unstable. The wormhole solutions, with their negative-density stuff, eliminate this difficulty because only a small fraction of the wave falls down a wormhole mouth each time it goes back in time. But if one has only positive-energy-density stuff, instability is a problem. In our time machine at the beginning of the Universe, however, you are *exiting* the time machine, and that is stable. If one disturbs the solution with a wave, the wave circulates clockwise around the loop. When it returns, because of the expansion of the branch, it has a wavelength 535 times longer than it did originally, and an energy only one 535th as great. Each time it goes around the loop, it comes back with an energy decreased by the same amount; therefore, even though the wave can circle an infinite number of times around the loop, it only causes a finite total buildup of energy, which does not greatly perturb the geometry. Thus, leaving a time machine is easier than entering one.

In our model, the Universe expands forever, so the Cauchy horizon extends forever in the future with no singularities on it. We escape Tipler's theorem because, although our model has some geodesic curves that do not extend to infinite length in the past, this occurs in a vacuum region without any real particles. Normally we might expect such curves to start either at a singularity, as Tipler's theorem assumes, or at a boundary such as the waist of de Sitter space in the tunneling solution. But in our case, such curves simply circle the time loop an infinite number of times. Particles circulating along such curves or photons going to the past along them could cause an infinite blowup of energy density in such a region. In our model, however, this occurs in a vacuum region where there are no real particles and where no photons are emitted toward the past, so this may not be a problem. Rather than have spacetime start with a singularity or a boundary condition, our model with a time loop establishes a periodic boundary condition at the start. (If a chaotic inflation model with all geodesic curves extending to infinite length in the past were shown to exist—its quantum fluctuations thereby violating some of the assumptions in Tipler's theorem—then we might make a model of our type with that property as well, by simply making one of the baby universes loop around and become the trunk.)

Thus, having a time machine appears easiest at the beginning of the Universe, when stability considerations are favorable and singularities are easiest to avoid. Li-Xin Li and I thought this very interesting, for that is just where we would need to have a time machine to enable the Universe to be its own mother.

191 *before Kip Thorne's work:* Barrow had written in 1986, "Some cosmological boundary conditions may be necessary either at an initial singularity or at past infinity (the alternative—that all timelike and null [lightlike] geodesics are closed, perhaps with periods >> [much greater than] 10^{10} years is not appealing)." As Thorne discusses

in *Black Holes and Time Warps*, in 1967 Robert Geroch proved a theorem that one could construct a wormhole by a smooth, singularity-free twisting of spacetime, but only if a time machine were created. Thorne writes, "The universal reaction to Geroch's theorem, in 1967, was '*Surely* the laws of physics forbid time machines, and thereby they will prevent a wormhole from ever being constructed classically, that is, without tearing holes in space.'" After Thorne's 1988 work, people were more willing to consider solutions involving time travel.

192 *Trade past for future:* This is known as *charge-parity-time*, or *CPT, invariance.*

198 *disorder increases with time today:* We were able to show that if you turned Figure 27 upside down to present a series of collapsing horns with a loop of time at the end of the Universe, then a self-consistent solution would require advanced waves only. Observers in such a Universe would see a time loop in the future and all light waves going toward the past. They would therefore observe causes occurring after effects. With the low-entropy time loop at the end, the entropy arrow of time would be reversed as well. Of course, people would rename the future the "past," and the past the "future." They would then think they lived "after" the closed loop of time, just as we do. In fact, the direction "toward the future" simply means "away from the time loop." Causes are always closer to the loop of time than effects are. Otherwise, the model would not be self-consistent, as an acceptable solution must be.

5. REPORT FROM THE FUTURE

210 *between* ¹⁄₃₉*th and 39 times its past longevity:* These rather broad limits are designed to catch 95 percent of the cases. The formula makes a correct prediction whenever the future longevity falls anywhere between these two limits. Often it will fall within a narrower range inside these limits. Recall, from the Berlin Wall, that half the time we expect the future longevity to be between ¹⁄₃rd and 3 times the past longevity. Thus, in most cases, the end arrives long before the 95 percent confidence level upper limit is reached.

Homo sapiens: R. L. Cann, M. Stoneking, and A. C. Wilson in 1987 estimated the age of our species, *Homo sapiens* (back to mitochondrial Eve), to be 200,000 years, based on DNA studies. This is the age for our species that I have adopted. This is in approximate agreement with other estimates: including 250,000 years, Gould (1989), p. 45n; greater than 100,000 years, R. Caroll, *Vertebrate Paleontology and Evolution* (New York: Freeman, 1988), pp. 475–476; and greater than 150,000 years, C. B. Stringer, *Scientific American* 263 (1990): 98.

210 *mammal species:* The mean longevity of mammal species is 2 million years, and the distribution of these lifetimes is an exponential distribution; see S. M. Stanley, *Proceedings of the National Academy of Sciences* 72 (1975): 646. Using this actuarial data on mammal species, we can set 95 percent confidence limits for the future longevity of a random mammal species alive today: more than 50,000 years but less than 7.4 million years. These limits are remarkably similar to the 95 percent confidence limits for the future longevity of the human race— more than 5,100 years but less than 7.8 million years—which are based solely on our own past longevity as an intelligent species.

222 *If you arrive at a random time:* The rule of 39 would likewise have kept you off *Titanic*'s sister ship, *Britannic*, which sank on its sixth voyage after hitting a German mine, but it would have allowed you many trips aboard *Titanic*'s other sister ship, *Olympic*, which made 514 Atlantic crossings before retiring.

One person who ignored the rule of 39 was White Star Line employee Violet Jessop, who was aboard both the *Titanic* and *Britannic* when they sank, escaping death each time. Astronauts must routinely bypass the rule of 39.

I'm not claiming that the first 39 voyages of a vehicle are necessarily dangerous. If you buy a new car, your first 39 trips may well go uneventfully, for in the case of your car, you may experience all of its voyages, from its very first to its very last. However, if you arrive at a dock to board a ship, you will be sampling, presumably, one of its trips picked at random. Of course, a ship may simply retire without catastrophe before 39 outings (as in the case of *Apollo 11*). The rule merely helps keep you off a ship's last voyage. Indeed, a conservative person might want to avoid such a trip because one way for a last voyage to end is in catastrophe. Particularly unlucky or intrinsically dangerous ships are unlikely to amass a long track record, so the rule helps keep you off such vessels.

A long, successful track record is a good safety indicator. Commercial airplanes are generally very safe and typically complete many thousands of flights without incident before they are retired. If you arrive at the airport at a random time, you are not, therefore, very likely to find a plane on its last trip nor on one of its first 39 trips— so the rule of 39 would not often interfere with your flight plans. One time in China I was asked to fly on a plane described as a "very old, very reliable Russian plane." It had been flying for 18 years, so I boarded it, figuring that, despite its decrepit appearance, my flight was unlikely to be its last. In flight, I mused that the Copernican principle was the only thing holding it up!

225 *95 percent that we are in the middle 95 percent:* Can you avoid the conclusion that you are likely to be in neither the first 2.5 percent nor the last 2.5 percent of the chronological list of human beings by arguing that you occupy a special position on the list by virtue of being born into an epoch when the level of sophistication was great enough to know the Copernican formula? If you are over 12 years of age, more than 1.8 billion people have been born after you already, pushing you off the last 2.5 percent of the list. If you are an optimist and believe that civilization is only going upward from here, then all future humans should live in epochs sophisticated enough to know or re-derive the formula. In this case, as someone who lives in an epoch sophisticated enough to know the formula, your probability of being in the first 2.5 percent of all humans is less than 2.5 percent (because such observers will occupy all of the chronological list except for a segment at the beginning). For you to be reading a formula like mine, all that is required is that you live in an epoch when it is known. After all, you live in an epoch when Copernicus's work is known, but you were not around when it was discovered, and it could be so with my formula as well. If civilization collapses and we return to a hunter-gatherer society not sophisticated enough to know such formulas, then the population is likely to be small (on the order of a million), and our future longevity would probably be similar to that for other hominids (on the order of 2 million years or less), making the likely number of future humans on the order of 100 billion—again, less than 2.7 trillion. Being in the first 2.5 percent of all humans requires good luck at the 2.5 percent level in all these scenarios.

229 *both treatments should agree:* If we actually had prior actuar-

ial data on the total populations through time of extraterrestrial intelligent species throughout the universe, then we could weight these by population (your chance of being a member of a particular species is proportional to the population) to produce an expected distribution for the likely total number of members of *your* intelligent species—*Homo sapiens*. This weighted distribution might have some characteristic scale—whether 100 billion or 100 trillion. But since we have no actuarial data on extraterrestrial intelligent species, we have no idea of what this scale might be. Thus, following Jeffreys, we should treat each a priori order-of-magnitude estimate of the total number of humans as equally valid. In other words, the total number of humans through time is considered a priori equally likely to lie in each of the following intervals: 100 billion to 1 trillion, 1 trillion to 10 trillion, 10 trillion to 100 trillion, and so forth. These estimates are then revised according to Bayes's theorem upon learning that you are approximately the 70 billionth human born. As I showed, this treatment leads to the Copernican results, namely that there is a 95 percent chance that the number of future humans lies between 1.8 billion and 2.7 trillion. A good vague prior, like that of Jeffreys, should be usable by any intelligent observer. If they all were to use it, then you could take a poll to see how well it worked; its results should agree with the Copernican answer because 95 percent of all those intelligent observers should be in the middle 95 percent of the chronological list of the members of their intelligent species.

238 *population now larger than the median:* By the same token, your intelligent species is likely to have a longer total longevity than the median intelligent species, because most of the intelligent observers are likely to come from such longer-lived species. And you would likely be among them. Thus we are likely to be more successful than the median intelligent species both in terms of longevity and population. Still, you should expect a 95 percent chance that you are located in the middle 95 percent of human history—giving us a projected total longevity of between 205,000 and 8 million years. It's just that the median intelligent species is likely to have a longevity that is even less than ours.

ANNOTATED REFERENCES

Abbott, E. A. *Flatland.* 7th ed. New York: Dover, 1952. Charming 1880 novel set in a world with two dimensions of space and one dimension of time.

Albrecht, A., and P. Steinhardt. *Physical Review Letters* 48 (1982): 1220. Bubble universes in inflation.

Alcubierre, M. *Classical and Quantum Gravity* 11 (1994): L73. Warpdrive in general relativity.

Alpher, R. A., and R. Herman. *Nature* 162 (1948): 774. Predicted cosmic microwave background radiation at a temperature of 5 degrees Kelvin.

Asimov, I. *Asimov's Biographical Encyclopedia of Science and Technology.* Rev. ed. New York: Avon Books, 1972. Asimov singlehandedly compiled this terrific resource, covering more than a thousand scientists.

Bardeen, J., P. J. Steinhardt, and M. Turner. *Physical Review D* 28 (1983): 679. Calculated fluctuations in inflationary cosmology leading to galaxy and cluster formation.

Barrow, J. D. In *Gravitation in Astrophysics*, edited by B. Carter and J. Hartle. New York: Plenum, 1987. Quote on possibility of closed timelike curves in the universe.

Barrow, J. D., and F. J. Tipler. *The Anthropic Cosmological Principle.* Oxford: Clarendon Press, 1986. Implications of the idea that intelligent observers must be found in habitable locations.

Bekenstein, J. D. *Physical Review D* 11 (1975): 2072. Proved black holes have entropy (disorder).

Benford, G. *Timescape.* New York: Pocket Books, 1980. Nebula Award–winning novel about time travel based on the many-worlds theory of quantum mechanics.

Bienen, H. S., and N. van de Walle. *Of Time and Power*. Stanford: Stanford University Press, 1991.

Birrell, N. D., and P. C. W. Davies. *Quantum Fields in Curved Space*. Cambridge, England: Cambridge University Press, 1982. Discusses Rindler vacuum.

Borde, A., and A. Vilenkin. *International Journal of Modern Physics D* 5 (1996): 813. Showed that in the bubble universe model, the original inflationary state must have a beginning.

Born, M. *Einstein's Theory of Relativity*. New York: Dover, 1962. A great book. Spacetime diagrams explain why moving observers disagree on which events are simultaneous.

Boulware, D. G. *Physical Review D* 11 (1975): 1404. Boulware vacuum outside a neutron star.

———. *Physical Review D* 46 (1992): 4421. Jinn particles and quantum probabilities.

Boyer, R. H., and R. W. Lindquist. *Journal of Mathematical Physics* 8 (1967): 265. Like Carter, they explore interiors of rotating black holes.

Browne, M. W. *The New York Times*, June 1, 1993, pp. C1, C7. Discusses my Copernican predictions for human longevity.

Burger, D. *Sphereland*. Trans. by C. J. Rheinboldt. New York: Crowell, Apollo Editions, 1965. Flatlanders discover they are living on the surface of a sphere.

Canavezes, A., et al. *Monthly Notices of the Royal Astronomical Society* 297 (1998): 777. A sample of 15,000 galaxies showing a spongelike clustering geometry.

Cann, R. L., M. Stoneking, and A. C. Wilson. *Nature* 325 (1987): 31. Estimated the age of our species based on DNA studies.

Carlini, A., and I. D. Novikov. Preprint TIT/HEP-338/COSMO-75 (1996). Self-consistency in time travel.

Carroll, S. M., E. Farhi, and A. Guth. *Physical Review Letters* 68 (1992): 263; erratum, 68 (1992): 3368; CTP#2117 (1992). Found that circling my cosmic strings gave your spaceship a 360-degree rotation plus a kick in velocity.

Carter, B. In *Confrontation of Cosmological Theories with Observations*. Ed. by M. Longair. Dordrecht: Reidel, 1974. The anthropic principle.

————. *Physical Review* 141 (1966): 1242; *Physical Review* 174 (1968): 1559. The complete extension of the interior of an unperturbed rotating black hole, showing a region of time travel trapped inside. Carter and I have two particular research interests in common: general relativity solutions involving time travel and examining the future of the human race.

Cassidy, M. J. *Classical and Quantum Gravity* 14 (1997): 523. Hawking's student proves that a self-consistent quantum vacuum for Misner space, allowing time travel, exists.

Chaitin, G. J. *Complexity* 1 (1995): 26. Brief account of Gödel's incompleteness theorem.

Cohen, J. E. *How Many People Can the Earth Support?* New York: Norton, 1995. Considers expert estimates of this figure: the median estimate is 12 billion people.

Coleman, S., and F. de Luccia. *Physical Review D* 21 (1980): 3305. Proposed that a de Sitter quantum vacuum state should decay through the formation of bubbles.

Corry, L., J. Renn, and J. Stachel. *Science* 278 (1997): 1270. The Hilbert-Einstein priority dispute resolved in Einstein's favor.

Cutler, C. *Physical Review D* 45 (1992): 487. Cauchy horizons in my two-string spacetime.

De Bernardis, P., et al. *Nature* 404 (2000): 955. Cosmic microwave background data consistent with inflation and suggesting that the part of the universe we can see is approximately flat.

Deser, S., R. Jackiw, and G. 't Hooft. *Annals of Physics* 152 (1984): 220. General relativity in Flatland.

Deutsch, D., and M. Lockwood. "The Quantum Physics of Time Travel." *Scientific American* 270 (March 1994): 68. A popular account of Deutsch's view of time travel.

Dewdney, A. K. *The Planiverse: Computer Contact with a Two-Dimensional World.* New York: Copernicus Books, 2001. Sequel to *Flatland.*

Duane, D. *The Wounded Sky.* New York: Pocket Books, 1983. *Star Trek* novel; includes references to papers by me and Mr. Spock.

Eaton, J. P., and C. A. Haas. *Titanic: Triumph and Tragedy.* 2nd ed. New York: Norton, 1994. Mentions how the Vanderbilts stayed off the *Titanic.*

Einstein, A. *Sitzungsberichte der Deutschen Akademie der Wissenschaften zu Berlin, Klasse für Mathematik, Physik und Technik* 1915 (1915): 844. The equations of general relativity!

———. *Sitzungsber., Preuss. Akad. Wiss.* (1917): 142. Einstein's static universe.

Everett, A. E. *Physical Review D* 53 (1996): 7365. Warpdrive time travel.

Farhi, E., A. H. Guth, and J. Guven. *Nuclear Physics* B339 (1990): 417. How to create baby universes in the lab.

Ferris, T. *New Yorker*, July 12, 1999, p. 35. Discusses my Copernican predictions.

———. *The Whole Shebang: A State-of-the-Universe(s) Report.* New York: Simon & Schuster, 1997. Masterful, wide-ranging treatment of the current state of cosmology. The black Taj inspires Ferris as a metaphor for dark matter.

Feynman, R. *The Character of Physical Law.* Cambridge, MA: MIT Press, 1994. Includes discussion of chess and the laws of physics.

Ford, L. H., and T. A. Roman. gr-qc/9510071 (1995). Showed that negative-energy-density stuff propping open wormhole tunnels must be confined to a narrow layer.

Friedman, J., M. S. Morris, I. D. Novikov, F. Echeverria, G. Klinkhammer, K. S. Thorne, and U. Yurtsever. *Physical Review D* 42 (1990): 1915. The principle of self-consistency in time travel.

Friedman, J. L., N. J. Papastamatiou, and J. Z. Simon. *Physical Review D* 46 (1992): 4442, 4456. How the fact that Jinn particles causing probabilities not to add up to 100 percent may be handled in Feynman's approach to quantum mechanics.

Friedmann, A. *Z. Phys.* 10 (1922): 377; 21 (1924): 326. Big bang models.

Frolov, V. P., M. A. Markov, and V. F. Mukhanov. *Physical Review D* 41 (1990): 383. Proposes baby universes are born in black holes.

Gamow, G. *One Two Three . . . Infinity.* New York: Dover, 1947. A remarkable book.

———. *Physical Review* 74 (1948): 505; *Nature* 162 (1948): 680. The hot big bang.

———. *Phys.* 51 (1928): 204. Quantum tunneling and radioactivity.

Garriga, J., and A. Vilenkin. *Physical Review D* 57 (1998): 2230. Showing how, if the universe today is dominated by a cosmological constant, this will lead to formation of high-density baby bubble universes in the future.

Gatland, K. *The Illustrated Encyclopedia of Space Technology*. 2nd ed. New York: Orion Books, 1989. Describes von Braun's plans to put astronauts on Mars by 1982, plus other useful space-program facts.

Giddings, S., J. Abbott, and K. Kuchař. *General Relativity and Gravitation* 16 (1984): 751. General relativity in Flatland. Professor Kuchař taught me general relativity in graduate school, perhaps the best course I ever took.

Gödel, K. *Reviews of Modern Physics* 21 (1949): 447. Gödel's rotating universe, allowing time travel to the past.

Gold, T. *Nature* 256 (1975): 113. How a mother could make her baby age less by assembling a dense, spherical shell of matter around the baby's crib each night.

Golden, F. *Time*, December 31, 1999, p. 62. *Time* names Albert Einstein "person of the century." Includes a tribute by Stephen Hawking.

González-Díaz, P. F. *Physical Review D* 59 (1999): 123513. Stability of Gott and Li self-creating Universe.

Gott, J. R. "A Grim Reckoning." *New Scientist*, November 15, 1997, pp. 36–39. My predictions for the human race and what we might do to increase our survival chances.

———. *Astrophysical Journal* 187 (1974): 1. My paper on tachyons that Benford used in his novel *Timescape.*

———. *Astrophysical Journal* 288 (1985): 422. Exact solution for one cosmic string.

———. *Il Nuovo Cimento* 22B (1974): 49. Tachyons in general relativity wouldn't transmit energy or information faster than light over macroscopic distances.

———. In *Inner Space/Outer Space*, edited by E. W. Kolb, et al. Chicago: University of Chicago Press, 1986. I explain that my open bubble universes can be produced by the particle physics scenario proposed by Linde and Albrecht and Steinhardt.

———. *M.N.R.A.S.* 193 (1980): 153. Examines images in pasted-together universes.

———. *Nature* 295 (1982): 304. Open bubble universes in inflation.

———. *Nature* 363 (1993): 315–319. "Implications of the Copernican Principle for Our Future Prospects."

————. *Nature* 368 (1994): 108. How the Copernican principle is consistent with a Bayesian approach.

————. "Our Future in the Universe." In *Clusters, Lensing, and the Future of the Universe*, Astronomical Society of the Pacific Conference series, edited by V. Trimble and A. Reisenegger, vol. 88, 140. San Francisco: Astronomical Society of the Pacific, 1996. More applications of the Copernican principle.

————. *Physical Review Letters* 66 (1991): 1126. My cosmic string time machine.

————. "Will We Travel Back (or Forward) in Time?" *Time*, April 10, 2000, pp. 68–70. Discussion of time travel and the laws of physics, as part of *Time*'s series "Visions 21," on the twenty-first century.

Gott, J. R., and M. Alpert. *General Relativity and Gravitation* 16 (1984): 243. General relativity in Flatland.

Gott, J. R., and L-X. Li. *Physical Review D* 58 (1998): 023501. "Can the Universe Create Itself?" How a time loop at the beginning could allow the Universe to be its own mother.

Gott, J. R., A. Melott, and M. Dickinson. *Astrophysical Journal* 306 (1986): 341. Showing that a spongelike geometry of galaxy clustering should result from inflation.

Gott, J. R., J. Z. Simon, and M. Alpert. *General Relativity and Gravitation* 18 (1986): 1019. General relativity and electrodynamics in Flatland.

Gott, J. R., and T. S. Statler. *Physics Letters* 136B (1984): 157. We set limits on the formation rate of open bubble universes.

Gould, S. J. *Wonderful Life*. New York: Norton, 1989. How biological evolution is chaotic and unpredictable in detail.

Grant, J. D. E. *Physical Review D* 47 (1993): 2388. *Grant space* is a time-travel spacetime, whose region of time travel was shaped like the time-travel region of my two-string solution. This facilitated certain calculations in the string case.

Greene, B. *The Elegant Universe*. New York: Vintage Books, 1999. Explains superstring theory.

Guth, A. H. *The Inflationary Universe*. New York: Perseus Press, 1997. Guth's personal account of his discovery of inflation.

————. *Physical Review D* 23 (1981): 347. Inflation.

Harrison, E. R. *Quarterly Journal of the Royal Astronomical Society* 36 (1995): 193. Proposes our universe might have been created in a lab by a previous intelligent civilization.

Hartle, J. B., and S. W. Hawking. *Physical Review D* 13 (1976): 2188. Hartle-Hawking vacuum state outside a black hole.

————. *Physical Review D* 28 (1983): 2960. Universe tunneling from nothing.

Hawking, S. W. *A Brief History of Time.* New York: Bantam Books, 1988. Hawking describes, among other things, the tunneling-from-nothing solution.

————. *Nature* 248 (1974): 30; *Communications in Mathematical Physics* 43 (1975): 199. Hawking radiation from black holes.

————. *Physical Review D* 46 (1992): 603. Hawking's chronology protection conjecture.

Hawking, S. W., and R. Penrose. *Proceedings of the Royal Society of London* A314 (1970): 529. Some theorems showing that a singularity in the early universe is inevitable—barring quantum gravity effects, closed timelike curves, or inflation.

Headrick, M. P., and J. R. Gott. *Physical Review D* 50 (1994): 7244. Our take on the Carroll, Farhi, and Guth paper. Plus comments about the possibility of a time machine based on a cosmic string loop hidden inside a black hole.

Heinlein, R. "All You Zombies—." In *Road to Science Fiction: Vol. 3. From Heinlein to Here.* Ed. by James Gunn. Clarkson, CA: White Wolf Publishing, 1979. Heinlein's story dates from 1959.

Hiscock, W. A. *Physical Review D* 31 (1985): 3288. Exact solution for one cosmic string.

Hiscock, W. A., and D. A. Konkowski. *Physical Review D* 26 (1982): 1225. Calculated quantum vacuum state for Misner space.

Hofstadter, D. *Gödel, Escher, Bach.* New York: Basic Books, 1979. Discusses Gödel's incompleteness theorem in mathematics.

Holst, S., and H-J. Matschull. *Classical and Quantum Gravity* 16 (1999): 3095. A lower-dimensional example (Flatland) in which there is a negative-energy-density vacuum state throughout space and a time machine of my cosmic string type is hidden inside a black hole.

Hubble, E. *Proceedings of the National Academy of Sciences USA* 15 (1929): 168. Hubble discovers the expansion of the universe!

Jeffreys, H. *Theory of Probability.* Oxford: Clarendon Press, 1939. Proposes the idea of a vague prior in Bayesian statistics, which is consistent with the Copernican outlook.

Jones, F. C. *Physical Review D* 6 (1972): 2727. Tachyon motion.

Kaku, M. *Hyperspace.* New York: Doubleday, 1994. Kaluza-Klein theories explained, plus a spacetime diagram of Heinlein's "All You Zombies—."

Kanigel, R. *The Man Who Knew Infinity: A Life of the Genius Ramanujan.* New York: Scribner's, 1991.

Kundić, T., E. L. Turner, W. N. Colley, J. R. Gott, J. E. Rhoads, Y. Wang, L. E. Bergeron, K. A. Gloria, D. C. Long, S. Malhotra, and J. Wambsganss. *Astrophysical Journal* 482 (1997): 75. We measured a time delay of 417 days for quasar 0957+561A,B resolving a controversy about the length of the delay from earlier data.

Lamoreaux, S. K. *Physical Review Letters* 78 (1997): 5. Measured Casimir vacuum in the lab with plates (actually a sphere and a plate) separated by 0.6 to 6 micrometers. His results on the force between the plates agreed with theory to within 5 percent.

Landsberg, P. T., J. N. Dewynne, and C. P. Please. *Nature* 365 (1993): 384. They used my 95 percent confidence Copernican formula to predict (correctly, as it turned out) the future longevity of the Conservative government in Britain.

Lemaître, G. *Ann. Soc. Sci. Bruxelles A.* 53 (1933): 51. A big bang model ending with an accelerating expansion due to a cosmological constant.

Lemonick, M. D. *The Light at the Edge of the Universe.* New York: Villard Books/Random House, 1993. The personal and scientific saga of the COBE results, vividly capturing the reactions among cosmologists.

———. *Time,* May 13, 1991, p. 74. Discusses my cosmic string time machine.

Leslie, J. *The End of the World: The Science and Ethics of Human Extinction.* London: Routledge, 1996; *Bulletin of the Canadian Nuclear Society* 10(3) (1989): 10; *The Philosophical Quarterly* 40 (1990): 65; *Mind* 101.403 (1992): 521. Summarizes the logic behind the kinds of arguments that he and Brandon Carter and Holgar Nielsen and I have been making about the number of future human beings.

Li, L-X. *Physical Review D* 50 (1994): R6037. The paper Li-Xin Li sent me in his letter, placing a reflecting sphere between wormhole mouths to prevent a quantum vacuum blowup.

Li, L-X., and J. R. Gott. *Physical Review Letters* 80 (1998): 2980. Self-consistent quantum vacuum for Misner space, allowing time travel.

Lightman, A. *Einstein's Dreams*. New York: Pantheon Books, 1993. Imaginative look at time.

Lightman, A., W. H. Press, R. H. Price, and S. A. Teukolsky. *Problem Book in Relativity and Gravitation*. Princeton, NJ: Princeton University Press, 1975. Self-supporting mass-shell size limits, included among many other interesting problems.

Linde, A. *Particle Physics and Inflationary Cosmology*. Chur, Switzerland: Harwood Academic Publishers, 1990. Linde's excellent book on his theory of chaotic inflation, whereby quantum fluctuations cause baby universes to grow like branches on a tree.

———. *Physics Letters* 108B (1982): 389. Bubble universes in inflation.

———. *Physics Letters* 129B (1983): 177. Chaotic inflation.

———. *Physical Review D* 59 (1999): 023 503. Open bubble universes in the context of chaotic inflation.

Lord, W. *A Night to Remember*. New York: Bantam Books, 1955. Mrs. Albert Caldwell's conversation while boarding the *Titanic* comes from this riveting account.

Lossev, A., and I. D. Novikov. Nordita preprint 91/41 A, submitted to *Classical and Quantum Gravity* (1991). Jinn in time machines and how the solutions must be self-consistent.

Marder, L. *Proceedings of the Royal Society of London, Ser. A*. 252 (1959): 45. Exact solution to Einstein's equations corresponding to a cosmic string—but before cosmic strings were proposed!

Mather, J. C., et al. *Astrophysical Journal Letters* 354 (1990): L37. COBE satellite shows microwave background spectrum is accurately thermal as predicted by the hot big bang model.

McKay, C. W., J. Kastings, and O. Toon. "Making Mars Habitable." *Nature* 352 (1991): 489–496.

Mendelssohn, K. *The Riddle of the Pyramids*. New York: Praeger, 1974. Examines why the Egyptians built pyramids and why they stopped.

Misner, C. W. In *Relativity Theory and Astrophysics I: Relativity and Cosmology*, Lectures in Applied Mathematics, edited by J. Ehlers, vol. 8, 160. Providence: American Mathematical Society, 1967. Misner space.

Misner, C. W., K. S. Thorne, and J. A. Wheeler. *Gravitation*. San Francisco: Freeman, 1973. I learned general relativity in Professor

Kuchař's course from this 1,279-page black book in its galley-proof stage. Includes Einstein's quotation describing his feelings after deriving the equations of general relativity.

Morris, M. S., K. S. Thorne, and U. Yurtsever. *Physical Review Letters* 61 (1988): 1446. Wormholes as time machines.

Nahin, P. J. *Time Machines.* New York: American Institute of Physics, 1993. Excellent book on time travel in science and science fiction.

Nielsen, H. B. *Acta Physica Polonica* B20 (1989): 427. On future population.

Novikov, I. D. *The River of Time.* Cambridge, England: Cambridge University Press, 1998. Emphasizes principle of self-consistency in time travel, viewing the past as unchangeable. Optimistic about the possibility of time travel to the past.

———. *Sov. Phys. JEPT* 68 (1989): 439. Self-consistency in time travel.

O'Neill, G. K. *Physics Today* 27 (September 1974): 32. Space colonies.

Ori, A. *Physical Review Letters* 67 (1991): 789; 68 (1992): 2117; and 71 (1993): 2517. Results suggesting that if one creates singularities in the process of making a time machine, one might still survive to time travel.

Pais, A. *Subtle Is the Lord . . .* Oxford: Clarendon Press, 1982. The best biography of Einstein. It has all the great stories. I have relied on this for a number of biographical details.

Penzias, A., and R. W. Wilson. *Astrophysical Journal* 142 (1965): 419. Nobel Prize–winning paper discovering the cosmic microwave background.

Perlmutter, S., et al. *Astrophysical Journal* 517 (1999): 565. Supernovae data suggesting that the expansion of the universe is accelerating.

Pickover, C. A. *Time: A Traveler's Guide.* New York: Oxford University Press, 1998. A nice introductory book on time-travel physics.

Preston, R. *The Hot Zone.* New York: Anchor Books, 1995. Points out the dangers of killer viruses.

Ratra, B., and P. J. E. Peebles. *Astrophysical Journal Letters* 432 (1994): L5, and *Physical Review D* 52 (1994): 1837. They calculated growth of structure from random quantum fluctuations in open bubble inflationary universes. M. Bucher, A. S. Goldhaber, and N. Turok,

Physical Review D 52 (1995): 3314, 5538, and K. Yamamoto, M. Sasaki, and T. Tanaka, *Astrophysical Journal* 455 (1995): 412, have continued these investigations.

Riess, A. G., et al. *Astrophysical Journal* 116 (1998): 1009. Supernovae data suggesting that the expansion of the universe is accelerating.

The Rig Veda. Translated by Wendy Doniger O'Flaherty. Harmondsworth, England: Penguin, 1981.

Sagan, C. *Broca's Brain.* New York: Random House, 1974. Discusses "Gott and the Turtles."

Schwarzschild, K. *Sitzungsberichte der Deutschen Akademie der Wissenschaften zu Berlin, Klasse für Mathematik, Physik und Technik* 1916 (1916): 189. When extended, this proved to be the black hole solution. The author died soon after writing this paper. His son, Martin Schwarzschild, was one of my mentors at Princeton.

Simon, J. Z. "The Physics of Time Travel." *Physics World* 7 (December 1994): 27–33. Notes that time machines with time loops of 5×10^{-44} seconds are the hardest to rule out. This is the type Li-Xin Li and I have proposed for explaining the beginning of the universe.

Smolin, L. *The Life of the Cosmos.* Oxford: Oxford University Press, 1997. How baby universes born in black holes could cause evolution in the physical constants to favor production of black holes.

Smoot, G. F., et al. *Astrophysical Journal* 420 (1992): 439. COBE satellite shows cosmic microwave background fluctuations consistent with inflation.

Stanley, S. M. *Proceedings of the National Academy of Sciences* 72 (1975): 646. Mean longevity of mammal species: 2 million years.

Staruszkiewicz, A. *Acta Physica Polonica* 24 (1963): 734. Point masses in Flatland.

Taylor, E. F., and J. A. Wheeler. *Spacetime Physics.* San Francisco: W. H. Freeman, 1992. A great book on special relativity, full of spacetime diagrams. My discussion of the "scratch test" for comparing moving light clocks is adapted from an argument they present.

Thorne, K. S. *Black Holes and Time Warps.* New York: Norton, 1994. A great book on black hole physics and time travel using wormholes.

Tipler, F. J. *Physical Review Letters* 37 (1976): 879. Showed under some general conditions that trying to make a time machine in a finite region with only positive mass material would lead to singularities.

Tyson, N. de G., et al. *One Universe: At Home in the Cosmos.* New York: John Henry Press, 2000. Beautiful book on our universe.

Unruh, W. G. *Physical Review D* 14 (1976): 870. Unruh radiation.

van Stockum, W. J. *Proceedings of the Royal Society of Edinburgh* 57 (1937): 135. Tipler later realized that this solution to Einstein's equations, an infinite rotating cylinder, allowed time travel.

Vilenkin, A. *Physical Review D* 23 (1981): 852. Approximate solution for one cosmic string.

———. *Physics Letters* 117B (1982): 25. Universe tunneling from nothing.

Vogeley, M., C. Park, M. J. Geller, J. P. Huchra, and J. R. Gott. *Astrophysical Journal* 420 (1994): 525. One of a number of studies done by different groups showing there is a spongelike geometry of galaxy clustering—consistent with inflation.

Wells, H. G. *The Time Machine* (1895), reprinted in *The Complete Science Fiction Treasury of H. G. Wells.* New York: Avenel Books, 1978. Started it all!

Wheeler, J. A., and R. P. Feynman. *Reviews of Modern Physics* 17 (1945): 157. Theory on arrow of time.

Wheeler, J. A., with K. Ford. *Geons, Black Holes, and Quantum Foam.* New York: Norton, 1998. Wheeler's autobiography.

Wilson, E. O. In *Biodiversity,* edited by E. O. Wilson. Washington, D.C.: National Academic Press, 1986. Species longevities.

Zubrin, R. M. *The Case for Mars.* New York: Free Press, 1996.

Zubrin, R. M., and C. P. McKay. "A World for the Winning: The Exploration and Terraforming of Mars." *The Planetary Report* 12(5) (September/October 1992).

INDEX